SOIL VAPOR EXTRACTION TECHNOLOGY

SOIL VAPOR EXTRACTION TECHNOLOGY

by

Tom A. Pedersen
James T. Curtis

Camp Dresser & McKee Inc.
Cambridge, Massachusetts

NOYES DATA CORPORATION
Park Ridge, New Jersey, U.S.A.

Copyright © 1991 by Noyes Data Corporation
Library of Congress Catalog Card Number: 91-12465
ISBN: 0-8155-1284-8
ISSN: 0090-516X
Printed in the United States

Published in the United States of America by
Noyes Data Corporation
Mill Road, Park Ridge, New Jersey 07656

10 9 8 7 6 5 4 3 2 1

Library of Congress Cataloging-in-Publication Data

Pedersen, Tom A.
      Soil vapor extraction technology / by Tom A. Pedersen, James T.
   Curtis.
            p.   cm. -- (Pollution technology review, ISSN 0090-516X ; no.
   204)
      Includes bibliographical references and index.
      ISBN 0-8155-1284-8
      1. Hydrocarbons--Environmental aspects. 2. Soil vapor extraction.
   I. Curtis, James T. II. Title. III. Series.
   TD879.073P43   1991
   628.5'5--dc20
                                                91-12465
                                                  CIP

Foreword

This book contains an assessment of the state-of-the-art of soil vapor extraction (SVE) technology and a summary of an expert workshop on SVE sponsored by the USEPA. Current SVE technologies are identified as well as additional research needed in such areas as site characterization, pilot systems, full scale system design and operation, attainment of cleanup criteria, and closure monitoring.

Over two million underground storage tank (UST) systems located at 700,000 facilities exist nationwide. The USEPA has indicated that about 25% of existing UST systems fail tightness testing and may be leaking.

Soil vapor extraction or vacuum extraction is an accepted, cost-effective technique for removing volatile organic compounds (VOCs) and motor fuels from contaminated soil. The technology is known in the industry by various names, including soil vapor extraction, vacuum extraction, soil venting, aeration, in situ volatilization, and enhanced volatilization. The growing interest in this technology is due in part to its demonstrated effectiveness for removing volatile compounds, its relatively low cost, and the apparent "simplicity" of system design and operation.

The first part of the book covers six major SVE topics: principles of soil vapor behavior, site investigations, system design, system operation and maintenance, secondary emission controls, and cost. The Appendices contain papers representing the wide diversity of topics presented at the expert workshop: a survey of SVE technology, two case studies, results of modeling studies, discussion of soil gas surveys, evaluation of biodegradation that occurs concurrent with SVE, and a discussion of research developments in SVE.

The information in the book is from *Soil Vapor Extraction Technology: Reference Handbook,* prepared by Tom A. Pedersen and James T. Curtis of Camp Dresser & McKee Inc. for the U.S. Environmental Protection Agency, February 1991.

The table of contents is organized in such a way as to serve as a subject index and provides easy access to the information contained in the book.

Acknowledgments

The Soil Vapor Extraction Technology Workshop, from which much of the information compiled in this document was obtained, and the preparation of this report were completed under the direction of Technical Project Monitor Chi-Yuan Fan, P.E., of the USEPA Superfund Technology Demonstration Division, Risk Reduction Engineering Laboratory, Releases Control Branch in Edison, New Jersey.

This report was prepared for the U.S. Environmental Protection Agency (EPA) Office of Research and Development by Camp Dresser & McKee Inc. (CDM) under the direction of Tom A. Pedersen, CPSS, Project Manager for Work Assignment No. 3-09 under EPA Contract No. 68-03-3409 with CDM Federal Programs Corporation. The principal authors of this report were Tom A. Pedersen and James T. Curtis. Additional material was provided by Paul B. Blais, Kathleen G. Murphy, Katharine Sellers, Yvonne L. Unger, and Michael L. Whitehead. Technical review was provided by Warren Lyman, Ph.D. and William Glynn, P.E.

CDM would like to acknowledge the guidance and assistance provided by Anthony N. Tafuri, ORD's Project Officer; Chi-Yan Fan, ORD's Technical Project Monitor for this work assignment; and Robert L. Stenberg, Ph.D. of USEPA, Risk Reduction Engineering Laboratory, Cincinnati, Ohio, who provided a complete technical review of the final document.

Special thanks is extended to the following individuals who presented papers at the workshop that are included as appendices to this document:

Hasan Cirpili, Hydrosystems, Inc., Dunn Loring, VA
Joseph Danko, CH_2M Hill, Corvalis, OR
David W. DePaoli, Oak Ridge National Laboratory, Oak Ridge, TN
Captain Michael Elliott, Tyndall Air Force Base, FL
Robert E. Hinchee, Battelle, Columbus, OH
George E. Hoag, University of Connecticut, Storrs, CT
Neil J. Hutzler, Michigan Technological Univ., Houghton, MI
Paul C. Johnson, Shell Development Corporation, Houston, TX
Henry B. Kerfoot, Kerfoot & Associates, Las Vegas, NV
Michael C. Marley, VAPEX, Philadelphia, PA
Robert J. Mutch, Eckenfelder, Inc., Mahwah, NJ

The presentations by the following workshop attendees are also greatly appreciated:

Glenn Batchelder, Groundwater Technology, Norwood, MA
Bruce Bauman, American Petroleum Institute, Washington, DC
Richard A. Brown, Groundwater Technology, Inc., Mercerville, NJ
Paul de Percin, USEPA RREL, Cincinnati, OH

Paul Lurk, USATHAMA, Aberdeen, MD
Mark Johnson, Chas. T. Main, Boston, MA
James J. Malot, Terra Vac, San Juan, PR
Frederick Payne, Midwest Water Resources, Inc., Charlotte, MI

The contributions of the workshop participants listed below are also acknowledged:

Ronald J. Backsai, Envirosafe Technologies, Inc.
Michael Broder, New York University, New York, NY
John E. Brugger, USEPA RREL, Edison, NJ
Paul C. Chan, New Jersey Institute of Technology, Newark, NJ
Jim Ciriello, Terra Vac, Princeton, NJ
Joseph G. Cleary, CH$_2$MHill, Parsippany, NJ
Gordon Dean, Florida Dept. of Environment Regulation, Tallahasee, FL
Dominic C. DiGiulio, USEPA, RSKERL, Ada, OK
Joseph C. Foglio, Environmental Applications, Inc., Waltham, MA
Iris Goodman, USEPA OUST, Washington, DC
Robert Hillger, USEPA RREL, Edison, NJ
Peter R. Jaffe, Princeton University, Princeton, NJ
Martin S. Johnson, Amoco Corporation, Tulsa, OK
William Librizzi, New Jersey Institute of Technology, Newark, NJ
Arthur E. Lord Jr., Drexel University, Philadelphia, PA
Joel W. Massman, Michigan Technological University, Houghton, MI
Emil Onuschak, Delaware Dept. Nat. Res. & Env. Control, New Castle, DE
Myron Rosenberg, Camp Dresser & McKee Inc., Boston, MA
John Schuring, New Jersey Institute of Technology, Newark, NJ
David E. Speed, IBM, Hopewell Junction, NY
Mary Stinson, USEPA RREL, Edison, NJ
David J. Wilson, Vanderbilt University, Nashville, TN

Special thanks are also extended to Gloria Veliz and Kathy Popplewell, who performed the word processing of the original manuscript; Valerie Zartarian, who contributed greatly to the production of the final report; and A. Russell Briggs, who prepared the original graphics, all of Camp Dresser & McKee Inc.

Abbreviations

ABNs   - Acid-Base Neutrals
API   - American Petroleum Institute
BTEX   - Benzene, Toluene, Ethylbenzene, and Xylenes
CEC   - Cation Exchange Capacity
CERCLA - Comprehensive Environmental Response, Compensation, and
          Liability Act of 1980 ("Superfund")
CFR   - Code of Federal Regulations
EM   - Electro-magnetics
EPA   - Environmental Protection Agency
FID   - Flame Ionization Detector
GAC   - Granular Activated Carbon
GC   - Gas Chromatograph
GPR   - Ground Penetrating Radar
LEL   - Lower Explosive Limit
NAPL   - Non-Aqueous Phase Liquid
PID   - Photoionization Detector
PVC   - Polyvinyl Chloride
RCB   - Releases Control Branch
RREL   - Risk Reduction Engineering Laboratory
SARA   - Superfund Amendments and Reauthorization Act
SCS   - Soil Conservation Service
SITE   - Superfund Innovative Technology Evaluation
SVE   - Soil Vapor Extraction
TDS   - Total Dissolved Solids
THA   - Total Hydrocarbon Analyzer
TPH   - Total Petroleum Hydrocarbons
UEL   - Upper Explosive Limit
USGS   - United States Geological Survey
UST   - Underground Storage Tank
VOCs   - Volatile Organic Compounds

NOTICE

The materials in this book were prepared as accounts of work sponsored by the U.S. Environmental Protection Agency. It has been subjected to the Agency's peer and administrative review, and has been approved for publication. On this basis the Publisher assumes no reponsibility nor liability for errors or any consequences arising from the use of the information contained herein.

Mention of trade names or commercial products does not constitute endorsement or recommendation for use by the Agency or the Publisher. Final determination of the suitability of any information or product for use contemplated by any user, and the manner of that use, is the sole responsibility of the user. The book is intended for information purposes only. The reader is warned that caution must always be exercised with potentially hazardous materials such as soil vapor, and expert advice should be obtained before implementation of processes involving soil vapor extraction technology.

All information pertaining to law and regulations is provided for background only. The reader must contact the appropriate legal sources and regulatory authorities for up-to-date regulatory requirements, and their interpretation and implementation.

Contents and Subject Index

APPENDIX F: DESIGN OF SOIL VAPOR EXTRACTION SYSTEMS – A SCIENTIFIC APPROACH

APPENDIX G: MODELING APPLICATIONS TO VAPOR EXTRACTION SYSTEMS

1. Introduction

Over two million underground storage tank (UST) systems located at 700,000 facilities exist nationwide. The U.S. Environmental Protection Agency (EPA) has indicated that about 25 percent of existing UST systems fail tightness testing and may be leaking (EPA, 1988a). Faced with the significant environmental impact to groundwater resources that releases from USTs pose, EPA, through their Office of Underground Storage Tanks (OUST), has embarked on an aggressive regulatory program under Subtitle I of the Hazardous and Solid Waste Amendments of 1984 (PL 98-616). Section 9003 of Subtitle I required EPA to promulgate regulations, applicable to all owners and operators of UST systems, to protect human health and the environment. Section 9003(h) was added by the Superfund Amendments and Reauthorization Act of 1986 (SARA) to give EPA and the states, under cooperative agreements with EPA, authority to clean up releases from UST systems or to require owners and operators to do so.

On September 23, 1988, in response to Section 9003, EPA promulgated final rules (40 CFR 280) on the technical requirements for underground storage tank systems (EPA, 1988a). The regulations describe a two-stage process for cleaning up petroleum or hazardous substances released from UST systems. The first stage addresses corrective actions taken in response to confirmed releases of regulated substances to mitigate immediate dangers posed by the release. The second stage addresses long-term corrective action measures undertaken if the implementing agency determines that additional corrective action is required to protect human health and the environment.

Since most UST releases result in contaminated soil, the regulations deal in part with soil treatment technologies, including:

- Excavation with on-site or off-site treatment or disposal;

- Biological treatment;

- Soil flushing;

- Recovery of non-aqueous phase liquid (NAPL); and

- Soil vapor extraction.

Soil vapor extraction (SVE) or vacuum extraction is an accepted, cost-effective technique for removing volatile organic compounds (VOCs) and motor fuels from contaminated soil. This technology is known in the industry

1

by various names, including soil vapor extraction, vacuum extraction, soil venting, aeration, in situ volatilization, and enhanced volatilization. In this report the term soil vapor extraction (SVE) is used. Vapor extraction systems have many advantages that make this technology applicable to a broad spectrum of sites:

- SVE is an in situ technology that can be implemented with a minimum of site disturbance. In many cases, normal business operations may continue throughout the cleanup period.

- SVE has potential for treating large volumes of soil at reasonable costs, in comparison to other available technologies.

- SVE systems are relatively easy to install and use standard, readily-available equipment. This allows for rapid mobilization and implementation of remedial activities.

- SVE effectively reduces the concentration of volatile organic contaminants in the vadose zone, which in turn reduces the potential for further transport of contaminants due to vapor migration and infiltrating precipitation.

- SVE can serve as an integral component of a complete remedial program, which may include groundwater extraction and treatment.

- Discharge vapor treatment options allow design flexibility required to satisfy site specific air discharge regulations.

The technical information available for designing, constructing and operating an extraction system is largely held by the SVE technology developers. Furthermore, it appears that engineering practices employed by vendors are based in large part on each developer's experiences. Therefore, an attempt has been made in this document to provide SVE technology design and operational information that would be of use in ensuring the appropriate and cost-effective application of these technologies.

The ease with which SVE systems can be installed and operated to achieve removal of volatile organic contaminants from the subsurface environment belies the complexity of vapor behavior in site specific subsurface settings. Some SVE users may be able to achieve significant VOC removal even without an in-depth understanding of subsurface vapor behavior. Nevertheless, further elucidation of the means by which SVE systems operate should lead to enhancement of system efficiency and increased confidence in attainment of clean-up standards.

The impacts resulting from past UST product releases, as well as those that will inevitably occur due to system failures and accidental discharges, present a significant remediation challenge. The EPA Office of Research and Development (ORD) is currently undertaking research into the applicability of a number of approaches for dealing with the contaminants released from UST systems. Soil vapor extraction is one such corrective action alternative that has been demonstrated effective in removing volatile contaminants from unsaturated soils.

One of the goals of this project was to develop technical assistance documents to assist in establishing design and operational parameters for vapor extraction systems, and to evaluate the effectiveness of the technology for removing the major gasoline constituents from contaminated subsurface zones. Parameters such as the layout and spacing of extraction and injection wells, induced vacuum/extracted flow relationships, transmission of the induced vacuum, vacuum pump capacity, air flow paths, and the effect of soil conditions (air permeability, temperature, moisture content, contaminant concentration, etc.) on system applicability are being considered and evaluated as part of these efforts.

Although SVE has become a widely used technology, few guidelines exist for the optimal design, installation, and operation of soil vapor extraction systems. The paucity of standardized design and operational criteria is due in part to the limitations posed by site specific factors including:

- contaminant characteristics and degree of weathering;

- extent of contamination;

- soil characteristics and stratigraphy;

- depth to groundwater;

- emission control requirements; and

- soil cleanup criteria.

ORD recognizes that SVE technology is being aggressively marketed by vendors and that data on the efficacy of these systems are sometimes not available because of the proprietary nature of the technology. Likewise, the evaluation of a site specific system application may not provide data that can be extrapolated to other sites due to variability in soils, geology, contaminant characteristics and SVE technologies. However, significant strides have been made recently in computer modeling of vapor phase transport of contaminants in unsaturated soils. In addition, laboratory soil column studies have yielded insight into the migration of contaminants released from leaking underground storage tanks. The synthesis of the results of computer modeling, laboratory studies and field data in a comprehensive, yet concise manner, and the distribution of these results, would afford greater uniformity in the evaluation and application of SVE. In turn, this should provide greater confidence in the use of the technology for remediation of contaminated soils where preliminary site conditions indicate a high probability of success.

WORKSHOP

The initial step in developing this document was the convening of SVE experts for a two day workshop in Edison, New Jersey on June 28 and 29, 1989.

Data on case studies, design factors and site evaluation approaches were presented and discussions were held to disseminate technology and to determine the most appropriate course of action for ORD with regard to technology dissemination and future SVE research. The data and information collected during the workshop are being used to develop an understanding of SVE technology and its limitations, while making this information available to individuals involved in site remediation. Specific issues suggested to ORD for future investigation include:

- Preparation of a technical assistance document that could be used to assess whether SVE technology is appropriate for implementation at specific UST sites. [Assessing UST Corrective Action Technologies: Site Assessment and Selection of Unsaturated Zone Treatment Technologies (EPA/600/2-90/011, 1990) has since been prepared by RREL-RCB];

- Preparation of design and operation guidance that would provide technical procedures for the application and evaluation of SVE technology;

- Development of a site screening procedure that could be used to assess whether SVE technology would be a viable option for site specific remediation;

- Performance of site screening procedures and field testing to assess SVE usefulness;

- Generation of data that could be used to estimate the site specific cost for site evaluation, screening, design and operation of SVE systems.

The following section addresses more fully the research and development ideas discussed during the workshop.

RESEARCH AND DEVELOPMENT

The ORD staff organized the SVE workshop reported on herein to develop a picture of the current state-of-the-art of SVE and to identify areas where additional research is needed. This section provides a brief synopsis of the thoughts expressed during the workshop on research and development needs. The experts generally agreed that ORD should undertake additional studies in the following three areas:

- Basic Research

- Applied Research and Development

- Regulatory Support

Basic Research

Workshop participants agreed that basic research into the behavior of contaminants in the subsurface environment is warranted. The type of research projects that should be given consideration are those that are predicated on the scientific method. Development of experimental designs that provide for generation of data to which statistical confidence limits and values can be assigned are necessary to allow for further clarification of soil vapor functions. Scholarly basic research serves as the foundation on which engineering advancements are built. The soundness of the engineering applications rests on the basic research finding. Failure of the engineering structure is often the result of faults in basic research.

Basic research in the following areas was recommended by workshop participants:

- Interaction of immiscible phase liquids in the capillary zone with unsaturated zone infiltration and saturated zone transport.

- Correlation of soil vapor and soil contaminant concentrations.

- Contaminant fate and transport in the unsaturated zone.

- Enhancement of biotreatment by SVE.

- Refinement of mathematical models to address complex soil environments.

- Identification of removal mechanisms mediating SVE clean-up.

- Factors affecting contaminant diffusion and retention in soils.

- Moisture content impacts on effective porosity and contaminant migration.

- Pulsed venting and desorption kinetics.

- The influence of micromorphological features on vapor migration.

- Ganglion formation and residual characterization.

Applied Research and Development

Applied research serves as a test of the findings of basic research. Field verification can be used to identify the limitations of extrapolating data obtained under controlled laboratory conditions to complex field situations. Applied research is undertaken not to refute theoretical findings of basic research but rather to obtain data that can be used to define the practical limitations of a system.

Applied research may also yield information on techniques that could potentially be applicable to SVE systems. The recent advancement in optical fiber technology provides an example of such an application. Fiber optics are gaining acceptance for use in analytical applications. Use of fiber optics in soil gas survey and monitoring of SVE systems is one area where applied research and development could yield advancement in engineering of SVE systems. Another example is the use of high-frequency radio waves for soil decontamination (Dev and Downey, 1988).

Most SVE vendors attending the workshop were adamant in their opinion that EPA should not be in the "business" of developing technologies. The development of technologies should be left to private enterprise, according to the vendors. They felt that divulging proprietary information would not be beneficial for advancement of SVE technology. Although it was recognized that some firms currently offering SVE services may not be cognizant of all the intricacies of soil environment or contaminants, or the limitations of the systems, the general consensus of the vendors was that the private sector should be allowed to advance the technology. In a free market, they contend, the SVE vendor offering the "better mousetrap" would be in the best position to advance the state-of-the-art.

Field performance evaluation of SVE has been undertaken at several sites with varying success. In some cases field tests have resulted in data that have not yielded meaningful conclusions regarding the system's effectiveness. These shortcomings may be due to the limited understanding of the fundamentals of vapor behavior and limitations of field and laboratory instrumentation and analytical techniques. The design and operation of adequately characterized evaluations would allow results to be extended to better understand SVE behavior at all sites.

During the workshop roundtable discussion it was noted that field demonstrations have in some cases preceded the basic research needed to adequately design the field tests. It was suggested that demonstration projects that yield inconclusive results may actually hamper future SVE efforts. The vendors who took part in the discussions were especially sensitive to the ramifications resulting from poorly designed field demonstration projects that could adversely influence regulatory and public perceptions about SVE technology. To forestall generation of faulty data, field demonstration projects should be undertaken only when the techniques and methods used will yield valid data.

Based on the discussions held during the workshop it was concluded that SVE applied research and development efforts should be concentrated in the following areas:

- Development of standardized SVE system monitoring and data interpretation approaches.

- Definition of temporal variations in overall extracted vapor quality, as well as the relative proportions of individual constituents.

- Determination of residual levels of contaminants in the soil at the conclusion of SVE.

- Determination of the relationship between extracted gas flow and the resultant zone of influence and upon cleanup times.

- Refinement of mathematical models to permit modeling of more complex soil conditions such as layered, heterogeneous soil systems.

- Evaluation of the effect of refrigerated condenser units on gas stream humidity and granular activated carbon life, and the quality of condensate from the refrigerated unit.

- Development of novel approaches and enhancements for SVE.

- Identification of variables that affect the rate of contaminant extraction and correlation of these variables with extraction rates.

- Evaluation of SVE effectiveness for petroleum products and dense halogenated products.

- Development of strategies to maximize saturated zone dewatering in the vicinity of immiscible phase liquids.

The American Petroleum Institute (API) is currently sponsoring a number of research projects related to SVE that may provide information on these topics. API research currently underway includes:

- Field evaluation of SVE, including measurement of the residual hydrocarbon content in the soil prior to and following vapor extraction.

- Quantification of mass balance hydrocarbon losses from sand columns.

- Pilot scale evaluation of soil venting.

- Field scale evaluation of SVE systems and venting well configuration.

- Quantification of the efficacy of subsurface venting in controlling vapor migration using multiple extraction vents.

Regulatory Support

Research to support regulatory development and implementation was identified as a major area of concern by SVE vendors, contractors, researchers, and regulators. The frustration of not being able to define an acceptable range of values for site clean up or even to define the parameter to be measured, was evident in the discussions during the workshop. It was agreed, despite the complexity of risk assessment and site specific conditions, that efforts should be made to standardize techniques for measuring contaminant concentrations in soil vapors and in soils that could be used as guidance by state regulators in setting standards. The lack of any clear guidance has led to the development of widely divergent cleanup standards. The areas requiring emphasis include:

- Establishment of risk based cleanup criteria for contaminants in soils under differing conditions.

- Standardization of soil gas sampling techniques.

- Standardization of laboratory analytical techniques for soil petroleum contaminant quantification.

- Guidance on acceptable sampling frequencies or statistical confidence intervals and limits for determining attainment of cleanup criteria and standardization of data reporting units.

REGULATORY CONSIDERATIONS

An important aspect of SVE design for each site is establishing a cleanup goal and a protocol that will be applied to ensure attainment of that goal, prior to commencement of the remediation effort. The clean up goal serves initially to guide the selection of the most appropriate remedial method and will also signal when site remediation has been achieved. This section provides an overview of the regulatory climate for both soil cleanup and air discharge.

Soil Criteria

Federal UST regulations (40 CFR 280) do not include specific soil cleanup standards; however, cleanup standards are suggested. A brief synopsis of the relevant subparts of the federal regulations governing UST corrective action plans follows.

Corrective Action Plan (40 CFR 280.66) --

(a)   Upon review of information from 40 CFR 280.61 - 280.63, the agency may require owners and operators to submit additional information or to develop a corrective action plan.

(b)   Plan must provide for adequate protection of human health, safety and environmental protection, and should include:

- characteristics of released substance, including toxicity and mobility

- hydrogeologic characteristics of facility and area

- proximity, quality, and present and future uses of nearby surface and ground water

- potential effects of residual contamination on surface and ground water

- exposure assessment

- any information compiled during fulfillment of these requirements

(c)   Upon approval, implement plan; monitor, evaluate, and report results to the satisfaction of the implementing agency

(d)   Corrective action may begin prior to approval if:

- notify agency of intention to begin cleanup

- comply with agency conditions

- incorporate these measures into corrective action plan submitted to agency for approval.

Public Participation (40 CFR 280.67)

(a)   If corrective action plan is required, the implementing agency must notify affected public

(b)   Agency must make information and decisions about corrective action plan available to the public

(c)   Before corrective action plan approval, agency may hold a public meeting to consider comments if sufficient interest exists

(d)   If corrective action plan has not achieved target, and termination of the plan is being considered, agency must notify the public.

Soil cleanup targets are determined on a site specific basis based upon state and local regulations and guidelines. Site specific criteria may be influenced by the location of the site as it affects human health and the environment. Remediation guidelines are generally set based on the risk associated with the contamination of public and private drinking water supplies or other exposure pathways (e.g., vapor exposure). Factors that are considered in the establishment of clean-up standards include: location of drinking water supplies; public utilities (such as sewers) in the area that may provide a pathway for future migration, and nearby buildings in the area that may be at risk of an explosion.

The determination of when a site remediation system is to be shut down and the site considered "clean" may require a site specific risk assessment. The following procedures serve as a guide for development of the risk assessment (Hinchee et al., 1986):

- Site characterization - soil types, groundwater levels and flow direction, bedrock formations.

- Hazard identification - hazardous substances at the site that could pose a threat to human health and environment.

● Fate analysis - pathways through which these hazardous substance
  may migrate (i.e. ground water, surface water, direct, air, etc.)

● Exposure assessment - level and degree of exposure to the
  hazardous substances via the site of the receptors.

● Risk determination - Risk is generally reported as a value related to
  the number of individuals exposed or at risk. For example, a value of
  1E-6 indicates that one out of one million people exposed is at risk.

● Risk management - levels to which contamination is reduced to
  achieve "acceptable" risk. "Acceptable risk" is typically
  defined as falling between 1E-5 and 1E-7.

This risk assessment approach can be standardized on a computer or other
method to rank each pathway for "risk," from some minimum to some maximum.
The State of California uses its "LUFT" (Leaking Underground Fuel Tank) field
manual for this purpose (Daugherty, 1984). These risks are then weighted and
summed to form a hazard ranking, with high risk sites receiving more priority
and low ranking sites receiving lower cleanup priority. Since many more sites
exist than can be addressed immediately, prioritization is necessary, and a
risk based system seems most reasonable. Low risk sites may best be addressed
by limited initial response followed by a determination of the need for long
term remediation (Duchaine, 1986).

Appendix K provides a listing of soil cleanup criteria for 50 states and
the District of Columbia. This table was compiled through a telephone survey
conducted in July and August, 1989. The table lists the office in each state
that oversees soil cleanup, and the name and telephone number of a contact in
that office.

This Appendix also lists the allowed residual. Each state falls into
one of several groups: states that have no regulations; states that determine
the allowed residual on a site-by-site basis; and sites that have specific
attainment standards, usually based on total petroleum hydrocarbon (TPH)
residual concentration.

Eleven states have yet to adopt formal regulations: Alaska, Arkansas,
Colorado, Connecticut, Florida, Idaho, Iowa, Kentucky, Maine, Pennsylvania,
and Wyoming. Many states surveyed indicated that no specific standards have
been established but rather each site is treated on a case-by-case basis.
This approach, while more subjective, allows for prioritization of the sites
based upon perceived risks. Three states -- California, Oregon, and Virginia
-- use risk assessment models to prioritize sites based on a ranking system
that takes into account toxicity, exposure pathways, and other criteria.
These models have also accounted for remedial action costs.

Twenty two states have established specific criteria for allowable
residual in the soil. The most common standards are based on total petroleum
hydrocarbon (TPH). The allowable limits range from 1 ppm (Ohio) and 2 ppm
(Nebraska) to 500 ppm in Oklahoma. Ten states use 100 ppm TPH, making it the

most common attainment standard. Michigan and New Hampshire use total
volatile organics rather than TPH as a standard. Three states use specific
compounds as cleanup criteria. Illinois has allowed levels for BTEX: benzene
(5 ppb), toluene (2 ppm), ethylbenzene (680 ppb), and xylenes (1400 ppb). New
Jersey specifies levels for three compounds: benzene (0.07 mg/l), toluene
(14.4 mg/l), and dichloroethylene (0.1 mg/l). Texas uses 500 ppm of BTX as
the indicator of cleanup.

Air Discharge Criteria

The removal of contaminants from the subsurface and their subsequent
discharge into the atmosphere without treatment, while sometimes feasible and
legal, fails to adequately address the core issue. Prior to 1986, though,
emission of toxic air pollutants was only minimally regulated. In that year,
Section 112 of the Clean Air Act required EPA to establish national emission
standards for hazardous air pollutants (NESHAPS). Since that event, some
toxic pollutants have received NESHAPS. The Emergency Planning and Community
Right to Know Act of 1986 has further increased public awareness of air
toxics, which may lead to further and increasingly complex air discharge
standards (Stever, 1989).

Appendix L lists the air discharge standards for 50 states and the
District of Columbia. These data were compiled from a state-by-state
telephone survey conducted in July and August, 1989. Nearly half (24 of 51)
of the states have no statewide air discharge standards and rely on federal
standards. California has no statewide standards. Rather, each county has
established its own limits. Many states do not recognize SVE systems as major
discharge points and regulate these systems as general emission sources.
Since many of the general emission source laws were written primarily for
large sources like power plants, SVE systems (which are by comparison small
sources) sometimes do not require treatment under present regulations.

Nine states have not established specific discharge regulations but do
require permits for SVE systems. Of these nine, two states, Kansas and
Oregon, require permits only for sites that discharge more than 10 tons of
VOCs per year.

Seventeen states have discharge limits expressed on a mass per time
basis. The states have widely varying allowable limits, however; for example,
North Carolina allows up to 40 pounds per day, while the District of Columbia
allows only 1 pound per day to be discharged. Other states have compound
specific emission limits. Connecticut, for example, lists over one hundred
compounds with an allowable limit based on both an 8-hour average and a
30-minute average. Some states (e.g., Vermont) require that the vapor
concentration in the influent stream be reduced by up to 85 percent.

In summary, the states vary widely in their air emission regulations,
from little or no formal regulation to contaminant-specific mass discharge
rates. Some base their standards on the concentration at the nearest
receptor, while others treat each site on a case-by-case basis.

FORMAT

The main text of this report is an assessment of the state-of-the-art of soil vapor extraction technology.   It was written specifically for state and local regulators, agency staff, environmental managers, remedial contractors and consultants who desire a basic understanding of SVE principles, applicability, operation, and cost.

A general overview of the theoretical considerations applicable to soil vapor extraction is provided in Section 2. This section includes discussions of the effect on SVE of contaminant properties, including vapor pressure, solubility, Henry's law constant, boiling point, soil sorption coefficient, contaminant composition and weathering and soil properties such as structure, moisture content, texture, air permeability, and temperature. Section 2 also discusses gaseous flow in the subsurface environment, including the equations that govern subsurface vapor flow. Finally, several field methods of determining the soil's air permeability are presented.

Section 3 provides an overview of site investigation approaches that can be used to obtain data necessary to determine if vapor extraction is a viable remedial option and, if so, obtain critical design information. This section also includes references to field techniques and equipment used to evaluate the site specific feasibility of vapor extraction.

General design approaches, including the determination of the air permeability, well selection and system configuration, are described in Section 4. In addition, this section discusses the components that comprise an SVE system. The purpose of this section is to provide the reader with a qualitative analysis of the design procedure and the individual components to aid in the preliminary design of such a system.

Operation, maintenance, and monitoring of SVE systems are discussed in Section 5. This section also includes discussions of enhanced biotreatment due to SVE, clean up attainment determination, including new methods for measuring residuals; and other issues related to system operation.

Section 6 discusses emission control methods available to treat the extracted vapors. Discussions are included on activated carbon adsorption, thermal and catalytic incineration, internal combustion engines, packed bed thermal processors, biotreatment, and direct discharge to the atmosphere.

The costs related to SVE implementation and operation are discussed in Section 7. This section discusses costs for a site investigation, component-by-component capital costs for SVE equipment, costs for prepackaged units, and operations and monitoring costs for these systems.

Ten appendices contain selected papers presented at the workshop held in Edison, New Jersey on June 28 and 29, 1989. Papers reprinted here were selected as representative of the wide range of topics discussed. Appendix A is a review of existing SVE operations by N.E. Hutzler, B.E. Murphy, and J.S. Gierke. This section reports on various aspects of SVE operations, including number, type, and layout of wells, type of blower or pump used, emission

control units and additional operational information. The section will give a
reader a sound historical basis with which to view other sections.

In Appendix B, J. Danko discusses the applicability and limitations of
SVE operation. This paper describes the advantages of SVE and discusses, from
an engineering viewpoint, some practical observations and advice.

Appendix C contains a report by H.B. Kerfoot on the use of soil gas
surveys in the design of SVE systems. Soil gas surveys are frequently used
during the site investigation phase to help to delineate the extent of
contamination and determine the types and relative concentrations of compounds
in the ground. With this information, a judgment can often be rendered
regarding the applicability of SVE for that site.

Appendix D, by R.E. Hinchee, D.C. Downey, and R.N. Miller, discusses the
enhancement of biodegradation that accompanies the use of soil vapor
extraction.

P.C. Johnson, M.W. Kemblowski, J.D. Colthart, D.L. Byers, and C.C.
Stanley contribute "A Practical Approach to the Design, Operation, and
Monitoring of In Situ Soil Venting Systems" in Appendix E. This report
presents a structured logical approach that forms a "decision-tree" for
deciding if SVE is appropriate to be used and, if so, describes the steps to
be taken during system design, operation, and monitoring.

Appendix F contains a scientific approach to SVE design in a paper by
M.C. Marley, S.D. Richter, B.L. Cliff, and P.E. Nangeroni. This paper
describes, among other things, the use of a computer model to calibrate data
obtained during a field air permeability test.

L.R. Silka, H.D. Cirpili, and D.C. Jordan discuss in Appendix G modeling
of subsurface vapor flow and the applications of modeling to SVE.

D.W. DePaoli, S.E. Herbes, and M.G. Elliot describe, in Appendix H, the
performance of SVE at a jet fuel spill site in Utah. This paper contains
knowledge and experience gained during SVE operation, operational results, and
a discussion of various aspects of SVE.

Appendix I also contains actual case history results for an industrial
site that has contamination from several volatile organic and base neutral
compounds, in a report by R.D. Mutch, Jr., A.N. Clarke, D.J. Wilson, and P.D.
Mutch. This interim report focuses on the measured zone of influence of the
extraction well, the composition of the extracted gas and its changes with
time, the treatability of the extracted vapor by granular activated carbon,
temperature variations that occur in the system, and groundwater upwelling due
to the induced vacuum. The authors also describe the use of a mathematical
model in their work.

A report by G.E. Hoag in Appendix J comments on recent SVE research
developments and research needs. These discussions follow a summary of SVE
"research milestones".

Appendices K and L to this document contain responses to a state-by-state survey conducted in August, 1989 regarding the allowable soil residual and air discharge criteria.

2. Principles of Soil Vapor Behavior

Products released into the subsurface environment from UST systems are acted upon by numerous forces that influence the degree and rate at which they migrate from the source. The extent to which the released products partition into the vapor phase is dependent upon the characteristics of the product, the nature of the subsurface environment, and the elapsed time since the release occurred. The manner by which the released product behaves in the subsurface will have a significant bearing on whether soil vapor extraction could be an appropriate corrective action for the site under consideration.

Figure 1 is a nomograph that uses the soil's permeability to air flow, the contaminant vapor pressure, and the time since release to predict the likelihood of success of a soil vapor extraction system. To use the nomograph, start at the time since release and draw a horizontal line to the appropriate soil air permeability. At the match line, draw a straight line to the vapor pressure. The point where this line intersects the continuum indicates how likely SVE would be at a site with that set of conditions.

This section provides a theoretical overview of the factors that influence contaminant fate and vapor phase transport in the vadose zone. The basic principles that govern soil vapor behavior and transport are identified to provide a sound basis for decision making with regard to site investigations, pilot testing, system design, operation and monitoring.

This section consists of three subsections: contaminant characteristics, soil environment and vapor transport. Contaminant characteristics considered include the physical and chemical properties of the released compounds, the composition of major petroleum classes, and how those properties affect their behavior in soil. Characteristics of the soil environment include those properties that affect the fate and transport of the released products. The section on vapor transport integrates the effects of the contaminant characteristics and the soil environment to show how vapor movement depends on those characteristics. The discussion focuses on how the product and soil factors influence the applicability of SVE systems, and also discusses the equations governing vapor flow under the influence of a vacuum.

CONTAMINANT CHARACTERISTICS

The physical and chemical properties of the released product control, to great extent, the movement and ultimate fate of that product in the subsurface. A product's properties affect the distribution of the product

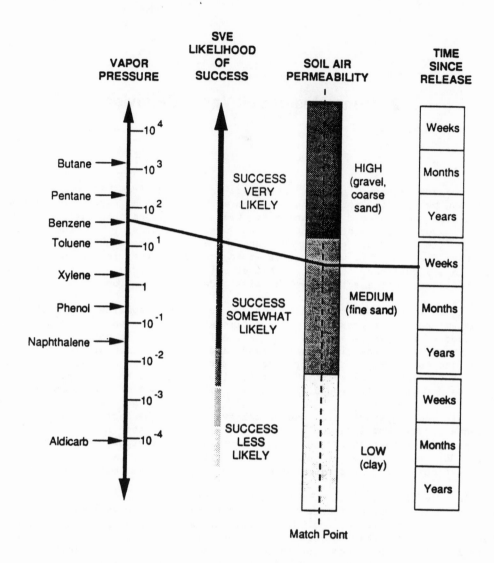

Figure 1. SVE Applicability Nomograph

among the four main phases in which the product may exist once released to soil (EPA, 1988a): (1) vapor, (2) dissolved in pore water, or (3) sorbed to a soil particles and colloids, and (4) as a non-aqueous phase liquid (NAPL) (Figure 2). The product's properties and other factors then affect the movement of each phase through the soil zone and how the compound's properties and movement change with time.

The distribution of a compound among the four phases can be described by several parameters. The degree to which a compound partitions into the vapor phase is described by that compound's vapor pressure and Henry's law constant. The soil sorption coefficient, K_d, describes the tendency of a compound to become adsorbed to soil. The solubility describes the degree to which a product will dissolve into water. The distribution of a product among the four phases varies with changes in site-specific conditions and will also change over time in response to weathering. It is important, however, to understand the mode in which a product exists prior to predicting transport of that product.

A product's volatility is directly associated with how much of that product will partition into the vapor phase in the soil gas. Highly volatile products, such as gasoline and certain chlorinated solvents, have a greater tendency to exist in the vapor phase than do other, less volatile products such as heating oil. The volatility of a product is perhaps the most important characteristic affecting the applicability of soil vapor extraction to that compound. The parameter that best describes a compound's volatility is its vapor pressure.

Vapor Pressure

All solids and liquids possess a vapor pressure, which-is a measure of the tendency of the substance to evaporate. Conceptually, vapor pressure can be considered analogous to the solubility of the material in air at a given temperature. Higher vapor pressures reflect an increased tendency to volatilize.

Vapor pressure is the force per unit area exerted by the vapor of the chemical in equilibrium with its pure solid or liquid form (Weast, 1981). For example, vapors from gasoline in a container will evaporate from the surface of the liquid gasoline and diffuse throughout the space occupied initially by air until an equilibrium is reached. The gasoline vapor that occupies the space above the liquid surface at equilibrium will exert a pressure on the container - the vapor pressure. The pressure within the container can be measured by the height to which it raises a column of mercury. Therefore vapor pressure is typically expressed in terms such as millimeters of mercury (mm Hg).

Chemicals with vapor pressures of less than 1E-7 mm Hg would generally not be expected to volatilize to a significant degree under ambient conditions (Dragun, 1988). Chemicals with vapor pressures of greater than 0.5 mm Hg would be expected to volatilize to a significant degree when released from an underground storage tank; these chemicals would be expected to respond to SVE technology (Bennedsen et al., 1985). Several of the major components of

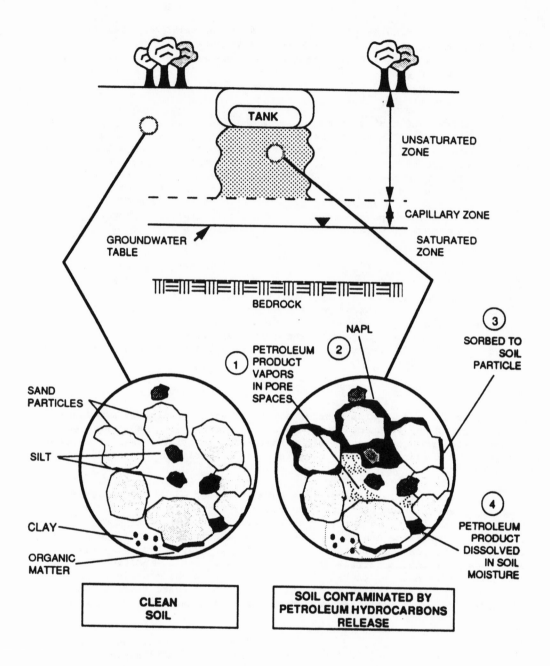

Figure 2. Unsaturated Zone Contaminant Phases

gasoline (as shown in Table 1) have vapor pressures greater than 0.5 mm Hg. These chemicals will have a much greater fraction of vapors in the subsurface environment. Thus, sites where release of these chemicals have occurred are likely candidates for remediation by SVE.

Temperature has a strong influence on the vapor pressure of a compound, with the vapor pressure increasing dramatically as temperatures increase. Jury et al. (1987) state that for most intermediate molecular weight organic compounds, vapor pressure increases three to four times for each 10 degree C increase in temperature. In general, vapor pressure may be approximated by:

$$VP(T) = A \exp(-B/T)$$

where T is the temperature in degrees Kelvin and A and B are constants characteristic of the substance. Vapor pressure must always be reported with the temperature at which the pressure was measured. Table 1 lists vapor pressures at a temperature of 20 degrees C.

The temperature of the subsurface environment therefore has an important influence on SVE technology. Research is currently underway to evaluate the effectiveness of steam injection or soil heating to increase volatilization and thereby increase the efficiency of soil vapor extraction systems (Hunt et al., 1986; Lord et al., 1987).

Water Solubility

The solubility controls the degree to which a product dissolves into ground water and pore water present in the vadose zone. Constituents present in petroleum products have widely varying solubilities. Table 1 shows that phenols, for example, are highly soluble while the heavier alkanes are only sparingly soluble.

Water solubility has an important impact on the movement of released product in the soil. Soluble products are likely to dissolve in infiltrating precipitation and move away from the source. Also, soluble products will move into the saturated zone and be transported away from the site, provided that the release is of significant volume. A fluctuating water table is another means of inducing groundwater contamination from soluble products.

Henry's Law

The volatilization of product dissolved in water is governed by Henry's law, which states that the concentration of a volatile chemical in a liquid is proportional to the partial pressure of the volatile chemical at low concentrations. The proportionality constant that relates concentration in solution to the partial pressure is known as Henry's law constant (K_H). Henry's constant may be reported as "dimensionless" or as $atm*m^3/mole$.

The Henry's law constant describes the relative tendency for a contaminant in solution to exist in the vapor phase. It is analogous to the vapor pressure, which describes the partitioning behavior between a pure

TABLE 1. CHEMICAL PROPERTIES OF HYDROCARBON CONSTITUENTS

Chemical Class	Representative Chemical	Liquid Density (g/cm.^{3}) @ 20°C	Henry's Law Constant (dim.)	Water solubility (mg/L) @ 25°C	Pure Vapor Pressure (mm Hg) @ 20°C	Vapor Density (g/m^3) @ 20°C	Soil Sorption Constant (Koc) (L/kg) @ 25°C
n-Alkanes							
C4	n-Butane	0.579	25.22	61.1	1560	4960	250
C5	n-Pentane	0.626	29.77	41.2	424	1670	320
C6	n-Hexane	0.659	36.61	12.5	121	570	600
C7	n-Heptane	0.684	44.60	2.68	35.6	195	1300
C8	n-Octane	0.703	52.00	0.66	10.5	65.6	2600
C9	n-Nonane	0.718	NA	0.122	3.2	22.4	5800
C10	n-Decane	0.730	NA	0.022	0.95	7.4	13000
Mono-aromatics							
C6	Benzene	0.885	0.11	1780	75.2	321	38
C7	Toluene	0.867	0.13	515	21.8	110	90
C8	m-Xylene	0.864	0.12	162	6.16	35.8	220
C8	Ethylbenzene	0.867	0.14	167	7.08	41.1	210
C9	1,3,5-Trimethylbenzene	0.865	0.09	72.6	1.73	11.4	390
C10	1,4-Diethylbenzene	0.862	0.19	15	0.697	5.12	1100
Phenols							
Phenol	Phenol	1.058	0.038	82000	0.529	2.72	110
C1-phenols	m-Cresol	1.027	0.044	23500	0.15	0.89	8.4
C2-phenols	2,4-Dimethylphenol	0.965	0.048	1600	0.058	0.39	NA
C3-phenols	2,4,6-Trimethylphenol	NA	NA	NA	0.012	0.09	NA
C4-phenols	m-Ethylphenol	1.037	NA	NA	0.08	0.53	NA
Indanol	Indanol	NA	NA	NA	0.014	0.1	NA
Di-aromatics	Naphthalene	1.025	NA	30	0.053	0.37	690

NOTE: NA - Not available

SOURCE: EPA, 1990.

substance (rather than a contaminant in solution) and its vapor phase.
Similar to vapor pressure, Henry's law constant is highly temperature
dependent and increases with increasing temperatures. Munz and Roberts (1987)
state that each rise of 10 degrees C in temperature corresponds to an increase
in the Henry's constant of 1.6 times.

The Henry's constant may be the more appropriate partitioning constant
outside of the free product zone, where product is likely to exist in solution
with pore water (Stephanatos, 1988). Henry's constant also applies in regards
to volatilization of contaminants from the groundwater into soil gas.

Table 1 lists the Henry's law constants for several common hydrocarbons
present in gasoline. The constituents listed in this table all have Henry's
law constants greater than 0.01 (dimensionless), the level at which compounds
have significant volatility making soil vapor extraction attractive (Danko,
1989; Hutzler et al., 1989a). Gasoline is particularly well-suited to SVE due
to its high composite volatility (for fresh gasoline, $K_H = 32$). Other
petroleum products, such as fuel oil No. 6, are less amenable than gasoline to
removal by SVE due to their lower volatility. However, researchers have
reported on the successful removal of petroleum products other than gasoline
through SVE; for example, DePaoli et al. (1989) report the successful removal
of JP-4 from a site in Utah.

Boiling Point

At a compound's boiling point, its vapor pressure equals the vapor
pressure of the atmosphere. At sea level the pressure exerted by the
atmosphere is 760 mm Hg. As elevation increases above sea level, atmospheric
pressure decreases rapidly; for example, at an elevation of 15,000 ft msl the
atmospheric pressure is 451 mm Hg. With a decrease in atmospheric pressure
comes a reduction in the boiling point. At a pressure of 451 mm Hg, for
example, water boils at a temperature of 86 degrees C, significantly less than
the boiling point at sea level (100 degrees C).

The relationship of boiling point to vapor pressure has important
implications with regard to SVE. This relationship describes a driving force
for movement of vapors from the liquid to the gaseous phase during vacuum
extraction. Inducing a vacuum in the soil causes the pressure in the soil
pore space to decrease. This, in effect, depresses the boiling point and
assists in driving the contaminant into the vapor phase. While the actual
magnitude of boiling point depression due to the induced vacuum is not always
a major factor in its volatilization, understanding this phenomenon should
provide an appreciation of the factors that might influence SVE operation.

Soil Sorption Coefficient

Sorption of contaminant liquids to soil particles and organic matter is
a very important factor: it controls the distribution of released products in
the soil zone and has a very strong effect on the movement of the product
through the vadose zone. In many cases the majority of the released product
may exist in the sorbed phase. Hinchee et al. (1987) state that for a 1000
gallon release at a "typical" site (as defined in the paper), the distribution

of the gasoline may be 962 gallons in the soil, 25 gallons in the soil gas, and 13 gallons in the groundwater. While these values are theoretical and dependent on many assumptions, the values do indicate the importance of considering the product that remains in the soil.

The sorption of a product to soil and organic matter is described by the contaminant's soil sorption coefficient, K_d. Values for K_d are not always readily available, so the more common octanol-water coefficient, K_{ow}, is often used as a surrogate for the soil sorption coefficient. The sorption coefficient describes the tendency for a product to sorb to the soil or organic matter. Table 1 lists the octanol-water coefficients for common hydrocarbon constituents. As this table shows, a strong relationship exists between the number of carbon atoms and the sorption coefficient, with larger molecules having a much greater tendency to sorb. This explains in part why compounds such as No. 6 fuel oil, which is high in these "heavy fractions", are very immobile in the subsurface (viscosity also plays an important part).

Contaminant Composition

Petroleum hydrocarbons are the most common products released from underground storage tanks (EPA, 1988a). Efforts to address the remediation of sites contaminated by releases from USTs often focus on the properties of the particular motor fuel that leaked, and on how the properties of that product affect the fate, mobility, and persistence in the ground. Petroleum hydrocarbons, however, are composed of many different compounds, each with different chemical and physical properties. Gasoline, for example, is a mixture of up to 200 compounds (CDM, 1986). Similarly, diesel fuel, heating oil, and other types of petroleum contain many different compounds in varying fractions. The characteristics of the bulk product (e.g., gasoline) will reflect the characteristics of the compounds that comprise the bulk product and the fraction of the whole that each constituent comprises.

Table 2 lists gasoline constituents, excluding additives, and the fraction each constituent comprises in a typical mixture, for both fresh and "weathered" gasoline. "Weathering" refers to the changes in the nature of a chemical mixture after its release into the environment. Weathering over time changes the product composition and will affect the ease with which that product is removed using SVE. Figure 3 shows the change in the composition of gasoline following weathering.

For example, each constituent of gasoline possesses a vapor pressure unique to that compound: its pure chemical vapor pressure. The vapor pressure of the gasoline is equal to the weighted average (by mole fraction) of the vapor pressures of all of the constituents, according to Raoult's law. Raoult's law states that the partial pressure of a volatile component (i) above a liquid mixture is given by:

$$p_i = p_i' * X_i$$

where p_i' is the vapor pressure of pure component i and X_i is the mole fraction in the liquid.

TABLE 2.   COMPOSITION OF FRESH AND WEATHERED GASOLINES

Compound Name	MW (g)	Fresh Gasoline	Weathered Gasoline
propane	44.1	0.0001	0.0000
isobutane	58.1	0.0122	0.0000
n-butane	58.1	0.0629	0.0000
trans-2-butene	56.1	0.0007	0.0000
cis-2-butene	56.1	0.0000	0.0000
3-methyl-1-butene	70.1	0.0006	0.0000
isopentane	72.2	0.1049	0.0069
1-pentene	70.1	0.0000	0.0005
2-methyl-1-butene	70.1	0.0000	0.0008
2-methyl-1,3-butadiene	68.1	0.0000	0.0000
n-pentane	72.2	0.0586	0.0095
trans-2-pentene	70.1	0.0000	0.0017
2-methytl-2-butene	70.1	0.0044	0.0021
3-methyl-1,2-butadiene	68.1	0.0000	0.0010
3,3-dimethyl-1-butene	84.2	0.0049	0.0000
cyclopentane	70.1	0.0000	0.0046
3-methyl-1-pentene	84.2	0.0000	0.0000
2,3-dimethylbutane	86.2	0.0730	0.0044
2-methylpentane	86.2	0.0273	0.0207
3-methylpentane	86.2	0.0000	0.0186
n-hexane	86.2	0.0283	0.0207
methylcyclopentane	84.2	0.0083	0.0234
2,2-dimethylpentane	100.2	0.0076	0.0064
benzene	78.1	0.0076	0.0021
cyclohexane	84.2	0.0000	0.0137
2,3-dimethylpentane	100.2	0.0390	0.0000
3-methylhexane	100.2	0.0000	0.0355
3-ethylpentane	100.2	0.0000	0.0000
n-heptane	100.2	0.0063	0.0447
2,2,4-trimethylpentane	114.2	0.0121	0.0503
methylcyclohexane	98.2	0.0000	0.0393
2,2-dimethylhexane	114.2	0.0055	0.0207
toluene	92.1	0.0550	0.0359
2,3,4-trimethylpentane	114.2	0.0121	0.0000
3-methylheptane	114.2	0.0000	0.0343
2-methylheptane	114.2	0.0155	0.0324
n-octane	114.2	0.0013	0.0300
2,4,4-trimethylhexane	128.3	0.0087	0.0034
2,2-dimethylheptane	128.3	0.0000	0.0226
ethylbenzene	106.2	0.0000	0.0130
p-xylene	106.2	0.0957	0.0151
m-xylene	106.2	0.0000	0.0376
3,3,4-trimethylhexane	128.3	0.0281	0.0056
o-xylene	106.2	0.0000	0.0274
2,2,4-trimethylheptane	142.3	0.0105	0.0012
n-nonane	128.3	0.0000	0.0382
3,3,5-trimethylheptane	142.3	0.0000	0.0000
n-propylbenzene	120.2	0.0841	0.0117
2,3,4-trimethylheptane	142.3	0.0000	0.0000
1,3,5-trimethylbenzene	120.2	0.0411	0.0493
1,2,4-ttrimethylbenzene	120.2	0.0213	0.0705
n-decane	142.3	0.0000	0.0140
methylpropylbenzene	134.2	0.0351	0.0170
dimethylethylbenzene	134.2	0.0307	0.0289
n-undecane	156.3	0.0000	0.0075
1,2,4,5,-tetramethylbenzene	134.2	0.0133	0.0056
1,2,3,4,-tetramethylbenzene	134.2	0.0129	0.0704
1,2,4-trimethyl-5-ethylbenzene	148.2	0.0405	0.0651
n-dodecane	170.3	0.0230	0.0000
naphthalene	128.2	0.0045	0.0076
n-hexylbenzene	162.3	0.0000	0.0147
methylnaphthalene	142.2	0.0023	0.0134
TOTAL		1.0000	1.0000

SOURCE: Johnson et al. (1989)

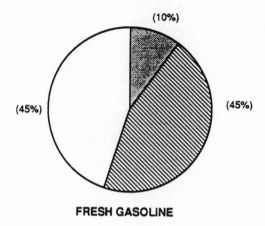

FRESH GASOLINE

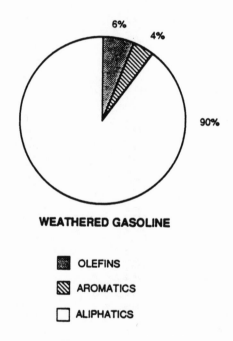

WEATHERED GASOLINE

OLEFINS

AROMATICS

ALIPHATICS

SOURCE: R.L. JOHNSON, 1989

Figure 3. Effect of Weathering on Gasoline Composition

Gasoline released into the subsurface will change composition as time passes because the more volatile components will partition into the vapor phase. This phenomenon will result in the remaining gasoline having a lower vapor pressure because it is relatively higher in the low-volatility compounds. Similarly, the more soluble components will preferentially dissolve versus the less soluble components, leaving the remaining gasoline less soluble overall.

The concept of weathering applies to SVE due to the effect SVE has on the composition of the residual product. At startup, the vapors removed by SVE are likely to be composed mainly of the more volatile, lighter-end fractions (molecules with five carbon atoms or fewer). After the system operates for some time, the extracted vapor will likely be depleted in the lighter-end fractions and will be composed mostly of heavier compounds. DePaoli et al. (1989) describe the occurrence of this phenomenon at a jet fuel spill site in Utah. Figure 4 shows how the composition of the extracted vapors changed as operation progressed. The figure shows that, initially, heavier molecules (C_7 and above) were a minor part of the extracted vapor, but became an increasingly large fraction as removal continued. This change is due to the loss of the more volatile compounds and an increase in the partial pressures of the less volatile compounds as their relative mole fractions in the liquid increase.

These concepts also have important implications with regard to SVE applicability as a corrective action alternative. SVE generally removes only the more volatile contaminants, often leaving in the soil those compounds that are less volatile. R.L. Johnson (1989) indicates that 20 to 35 percent of the product as total petroleum hydrocarbons (TPH) will remain in the soil. This residual will be composed predominantly of C_{10} and above hydrocarbons. These residual components may or may not be important from a risk standpoint, but it should be recognized that, depending on the standards for allowed residual, SVE may not remove all product constituents to an acceptably low level.

The air permeability of the soil incorporates the effects of several soil characteristics, the most important of which are soil structure, stratigraphy, air-filled porosity, particle size distribution, water content, residual saturation, and the presence or absence of macropores.

SOIL ENVIRONMENT

The soil environment, like the product characteristics, has a significant effect on the fate and transport of products released into the subsurface. Differences in the soil type, structure and stratigraphy, the particle density, and the porosity will affect the ease with which vapors will pass through the soil and will determine the air permeability of the soil zone. Other parameters, such as moisture content and organic matter content, will affect air movement through the soil zone and, more importantly, may retard the transport of contaminants in the soil gas by adsorption onto the soil or dissolution into pore water. Because soil properties have such an important effect on the movement of soil gas, the efficacy and design of soil vapor extraction systems are largely dependent on the soil properties.

Source: De Paoli et al.,1989

Figure 4. Variation of Hydrocarbon Distribution in Extracted Gas

Air Permeability

The soil's permeability to air flow is simply a measure of the ability of vapors to flow through porous media. It is analogous to the permeability to water flow in the saturated zone. The permeability of the soil to air flow is perhaps the single most important parameter with respect to the success of soil venting. It is a key parameter not only in deciding if SVE is a feasible remedial option, but also for the SVE system design (Johnson et al., 1989).

The air-filled porosity is a basic determinant of the volume available for vapor transport. Water content, or the percentage of the porosity filled with water, has a great effect on the pore space available for vapor transport, as discussed above. The air permeability is significantly influenced by the density and viscosity of the soil gas (Krishnayya et al., 1988). The density and viscosity are in turn affected by temperature. Table 3 shows the effect of temperature on water and air properties. The table developed from theoretical considerations, shows the ratio of hydraulic conductivity to air conductivity at various temperatures, and how this ratio increases with increasing temperatures. This ratio can be used to estimate the air permeability of a soil from its hydraulic conductivity.

Porosity

Vapor migration in the subsurface occurs principally through air filled pore spaces. Diffusion of gases through liquid filled pore spaces does occur, albeit at rates a thousand times less than that which occur through air filled pores. Factors influencing vapor migration include the air-filled porosity (which is affected by the water content) of a soil and the orientation of the soil pores.

TABLE 3. EFFECT OF TEMPERATURE ON WATER AND AIR PROPERTIES AND AIR CONDUCTIVITY

Temperature $^{\circ}C$	DYNAMIC VISCOSITY (kg/m.S)		DENSITY AT STANDARD ATMOSPHERIC PRESSURE (kg/m^3)		HEAD AT STANDARD ATMOSPHERIC PRESSURE (m)		RATIO OF HYDRAULIC CONDUCTIVITY TO AIR CONDUCTIVITY
	Water x 10^{-3}	Air x 10^{-5}	Water	Air	Water	Air	(K_r)
0	1.79	1.71	999.868	1.2930	10.333	7991	7.4
5	1.52	1.74	999.992	1.2697	10.332	8137	9.0
10	1.31	1.77	999.728	1.2472	10.335	8284	10.8
20	1.00	1.83	998.234	1.2047	10.350	8576	15.2
30	0.80	1.86	995.678	1.1649	10.377	8869	19.9
40	0.65	1.90	992.247	1.1200	10.413	9225	25.9

NOTES: 1) Standard Atmospheric Pressure = 10.332 kg/m^2
2) Ratio of Hydraulic Conductivity to Air Conductivity:
K_r = (viscosity of air/viscosity of water) (Density of water/Density of air)
3) Properties of water and air are from Weast (1970)
SOURCE: Krishnayya et al.,1988

Soil Structure

The soil type affects the fate and transport of products released from USTs and, therefore, the applicability of SVE. The porosity, or the fraction of voids in the soil, is important to SVE. Soils with higher porosities will allow a higher flow for the same induced vacuum. Michaels and Stinson (1989) report that porosity appears to be an important parameter, based on their SITE (Superfund Innovative Technology Evaluation) demonstration. At that site, highly permeable sands and impermeable clays (both with porosities of 40 to 50 percent) exhibited good removal rates via SVE. Porosity is related to the particle size distribution of the soil. Coarse-textured soil will generally have higher permeabilities than fine-textured soils.

Residual Saturation

Product migrating from an UST leak may coat the soil through which it passes. When the release ceases, the product may continue to migrate under the influence of gravity, soil matric potential and pressure gradients. That product which remains in the soil after the saturation front has passed and after free gravity drainage has occurred, is referred to as "residual saturation". The amount of product that is retained at residual saturation will vary depending on the soil characteristics and the product composition.

Jury and Ghodrati (1989) estimate that 5 to 10 percent of available pore space may be occupied by non-aqueous phase liquids (NAPL) after the wetting front moves through the soil. Hoag and Marley (1986) also report on the results of column tests showing how residual saturation varies with soil type and moisture content. Coarse particles held less product than fine particles at residual saturation. Soils that were wet originally were shown to retain from 20 to 30 percent less (for medium sand) up to 60 percent less (fine sand) product than soil that was originally dry. In general, residual saturation values ranged from 12 to 60 percent saturated.

NAPL remaining in the soil under these conditions is immiscible and not subject to significant migration in response to wetting fronts from infiltrating rainfall or induced wetting fronts. This factor tends to favor the use of SVE systems over in situ soil flushing options. SVE induces product to volatize into the vapor phase, where contaminants are far more mobile. Thus, otherwise immobile residual liquids are volatilized into the vapor phase, where they are more easily removed than as NAPL. Soil flushing attempts to mobilize liquid and sorbed contaminants through dissolution and a pressure gradient. Researchers focusing on petroleum residual saturation have noted that once the products have sorbed to soil, they are often extremely difficult to remove as a liquid.

Ganglia Formation

Soil structure and the physical arrangement of pore spaces in the subsurface influence the manner in which contaminants migrate from a source. Product released from an UST system will move through soil under the influence of gravity and other gradients to varying degrees and at differing rates depending on the soil characteristics. In a situation where the soil surrounding an UST is a uniform sand, one would expect migration at a greater rate than in a clayey soil, unless significant cracks or macropores exist in the clay. Migration of fluids in soils has been shown to predominate in macropores. The extraction of vapor from a soil would therefore be expected to occur initially from the larger continuous pores rather than from small continuous pores. As products migrate from a source they may collect in subsurface pools, cracks, or fissures. The isolated globules of product that may form in soil have been referred to as "ganglia" (Jury and Ghodrati, 1989). Ganglia may be especially prevalent in soils where immiscible petroleum products have been released as compared to sites where miscible products have leaked.

Subsurface Conduits

Products released from an UST do not generally flow through soils in a uniform wetting front. Flow will be diverted due to variations in soil structure and horizons. The presence of subsurface features that are less restrictive to flow may result in preferential flow in directions different than what would be anticipated from a review of localized flow patterns or soil strata. For example, subsurface utility conduits (such as electric, telephone, sewer, or water lines) may be bedded in gravel or materials more permeable than surrounding soils. Liquids and gases may migrate along these preferential flow paths for significant distances, whereas migration in the bulk soil mass may be limited. Because the majority of USTs have been installed in areas where subsurface disturbance has occurred, it is important to consider these features in assessing vapor migration. In general, any feature that favors liquid product transport would also serve as a preferential flow path for vapors.

Water Content

Water content of the soil has competing effects on the air permeability. The primary effect of pore water is to reduce the air-filled pore space of a soil. Stephanatos (1988) concludes that the movement of soil gas is reduced as water content increases due to the physical reduction in available air pathways. Figure 5 shows the relationship between air permeability and water permeability. Stonestrom and Rubin (1989) relate the air permeability to matric pressure and trapped air, and also show the reduced air permeability for higher matric pressure. This body of research indicates that SVE would be more successful at lower water contents since a greater percentage of the pore space is air-filled and available for vapor transport, and thus the induced air flow is greater for a given vacuum.

The water content also has a significant effect on the success of SVE through its effect on the sorption characteristics of organic compounds. Researchers have concluded that the reduction in air-filled pore space that occurs as water content increases serves to reduce volatilization (Farmer et al., 1980; Aurelius and Brown, 1987). Lighty et al. (1988) and Houston et al. (1989) show that the soil sorption coefficient is greater for drier soils; as water content increases, sorption of contaminants to soil decreases as water displaces contaminant molecules. Reible (1989) showed that electrostatic

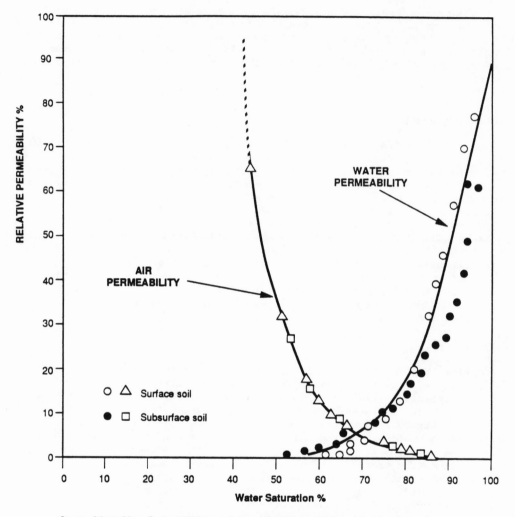

Source: Adapted from Corey, 1957 by permission of Soil Science Society of America, Inc.

Figure 5. Air and Water Permeability as a Function of Water Content

force increases for drier soils, leading to reduced volatilization from the dissolved phase into the soil gas. Figure 6 shows schematically the effect of varying soil moisture regimes on soil sorption. Davies (1989) reports that the soil sorption coefficient may be four orders of magnitude higher for dry soils than for wet soils.

The optimal soil moisture regime for SVE applications is a water content low enough to ensure adequate air permeability yet wet enough to reduce electrostatic sorption force. Davies (1989) states that the critical moisture regime for SVE applications is 94 to 98.5 percent relative humidity in the soil gas. Below this range volatiles are more tightly bound to soil and may not be as readily volatiled. In regions such as the southwestern United States, where ambient soil moisture is typically very low, the use of humidified air may be warranted if the soil moisture is low enough that adsorption forces become significant (Davies, 1989). Some researchers (Hunt et al., 1988; Lord et al., 1987) have investigated steam injection to volatilize contaminants, both those sorbed to soil and those that exist as non-aqueous phase liquid (NAPL) in the soil. The results show potential for this type of operation, although its applicability for wide scale use has not been proved.

Preferred Flow Paths

Preferred flow paths, such as macropores formed by cracks, root casts, or earthworms, are highly permeable, continuous voids that may transmit a significant quantity of the vapor or liquid through a soil (Levy and Germann, 1988). Researchers have shown that these macropores often control the rate of infiltrating water in the soil. This has major implications with regard to vapor extraction because significant quantities of flow may be realized in continuous macropores leading to the vapor extraction system. Because of this short circuiting phenomenon the contaminants adhering to soils through which limited air flow occurs may not be cleaned up as rapidly as would those particles through which a large volume of air flows. Removal of contaminants from soil particles in dead end pores (i.e., pores that are connected to other pores only at one end) would therefore be more closely related to diffusion effects. Removal of product from these dead end zones may not be enhanced by increased vacuum or pumping rates, but rather removal will be limited by diffusion from these zones to the more continuous flow paths. Diffusion of contaminant vapors occurs much more slowly than advection. Soils with lower air permeability values are therefore likely to be less amenable to soil vapor extraction or will at least require higher vacuums.

GASEOUS FLOW IN SUBSURFACE ENVIRONMENTS

Air flow rates and subsurface permeabilities have historically been used in the petroleum industry to estimate extraction of natural gas from oil- and gas-producing geological formations. The petroleum industry tests typically are conducted in the field, employ gas extraction wells, and are performed on confined subsurface strata. Gas producing strata are generally found at significant depth and have much higher natural gas pressures and temperatures than the situations encountered at UST sites where SVE technology could be applied. Soil gas permeability estimates have also been used in agricultural

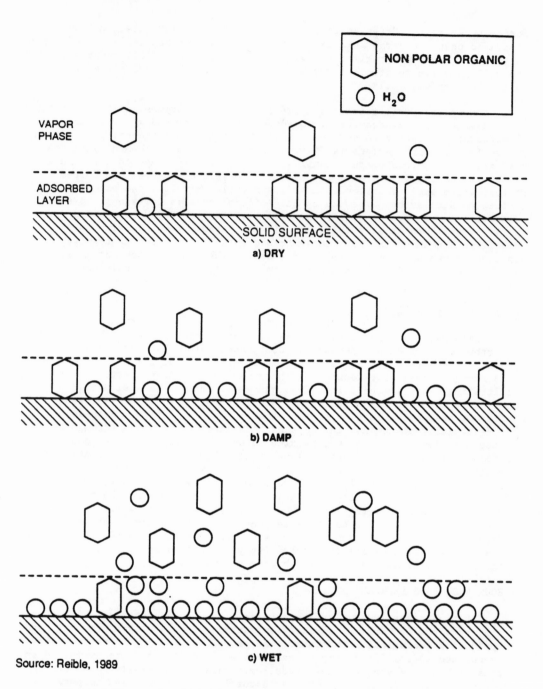

Source: Reible, 1989

Figure 6. Illustration Of VOC Adsorption Under Three Moisture Regimes

research to determine the amount of atmospheric nitrogen and oxygen available to plants in the root zone. Agricultural soil gas permeabilities are usually determined from laboratory measurements or in situ air injection tests.

While the situations explored by petroleum and agricultural researchers do not correspond exactly to most SVE situations, the fundamental vapor transport principles used by researchers in these fields do apply to transport of contaminant vapors for SVE theory and practice. This section provides a brief description of the basic principles controlling vapor flow in porous media, as well as available methods and techniques to measure air permeabilities.

Many factors influence contaminant vapor transport and diffusion. Table 4 lists several parameters that influence diffusion and loss to the atmosphere, including water content, adsorption site density, chemical concentration, temperature, and Henry's law constant.

Vapor Transport Fundamentals

Vapor flow through soil is dependent on soil characteristics, including porosity and permeability; gas properties, such as viscosity and density; and pressure gradients. Gas is a fluid and as such its flow rate through porous media is commonly characterized by Darcy's law. Darcy's law is valid for laminar, isothermal flow that is uniformly distributed across a given cross-sectional area. The general formulation of Darcy's law for saturated fluid flow in one dimension is:

$$Q = (kA/\mu)(dP/dm) \tag{1}$$

where:
  Q = flow rate (cm^3/sec)
  k = permeability (cm^2)
  A = cross-sectional area (cm^2)
  μ = viscosity (g/cm*sec)
  dP/dm = pressure gradient ($(g/cm*sec^2)$/cm)

The use of Darcy's law is restricted to laminar flow situations, where individual fluid particles are considered to flow parallel to the soil pore walls. Fluid flow is characterized by laminar or turbulent flow based upon its Reynolds number. Generally, laminar flow conditions exist when the Reynolds number (R) is less than about 2100:

$$R = \frac{V\rho d}{\mu} \tag{2}$$

where:
  R = Reynolds number (dimensionless)
  V = velocity of the fluid (cm/sec)
  ρ = density of the fluid (g/cm3)
  d = effective particle diameter (cm)
  μ = viscosity of the fluid (g/cm*sec)

TABLE 4. VAPOR DIFFUSION-PARAMETER RELATIONS IN
CHEMICAL TRANSPORT AND LOSS TO ATMOSPHERE

Parameter	Influence
Water content (Θ)	Decreases effective porosity for vapor flow; also, vapor diffusion decreases strongly with increasing water content. A frequently used model assumes that the soil vapor diffusion coefficient is proportional to $(\emptyset-\Theta)$ exp 3.33, where $\emptyset$ is porosity.
Adsorption site density	Adsorption decreases gaseous chemical concentration and decreases vapor diffusion. Most volatile organic chemicals are nonpolar and adsorb primarily to organic matter.
Chemical concentration	For chemicals whose vapor density is not saturated, increasing chemical concentration will increase vapor density and increase vapor diffusion. The increase may be greater than proportional if the chemical vapor adsorption isotherm is nonlinear.
Temperature	Increasing temperature significantly increases vapor density for a given amount of chemical in soil, thereby increasing vapor diffusion. However, soils typically remain at a constant temperature at a particular site. The vapor diffusion coefficient increases nonlinearly with increasing temperature, proportional to T exp 1.75 (Kelvin).
Henry's Constant K_H	Henry's constant (ratio of saturated vapor density to solubility) is an index of the partitioning of a chemical between dissolved and gaseous phases. Compounds with larger K_H values are more likely to move by vapor diffusion as opposed to liquid diffusion.

Source: Jury and Valentine, 1986

Johnson et al. (1990) developed vapor flow rate and air permeability estimation methods assuming laminar, steady-state vapor flow conditions for a soil venting well. The basic governing equations used to model this flow are the continuity equation and Darcy's law. The continuity equation states that the mass flow rate of a fluid through a given cross-sectional area remains constant with time. The forms of these equations that can be used in this analysis are (Johnson et al., 1990):

$$\frac{\partial(\varepsilon\rho m)}{\partial t} = -\nabla(\rho m\mu) \tag{3}$$

$$u = \left(\frac{-k}{\mu}\right)(\nabla p + \rho g) \tag{4}$$

where:

ε = vapor-filled void fraction $(0 < \varepsilon < 1)$
ρ = vapor density (g/cm^3)
m = stratum thickness (cm)
t = time (sec)
u = Darcian vapor flow velocity (cm/sec)
k = air permeability (cm^2)
μ = vapor viscosity (g/cm*sec)
g = gravitational acceleration (cm/sec^2)
∇ = gradient operator (1/cm)
P = vapor-phase pressure $(g/cm*sec^2)$

The gravity term in equation (4) is generally neglected due to the low density of the vapor. An expression for radial soil vapor flow to an extraction well is developed by substituting Darcy's law into the continuity equation and replacing the vapor density with a density-pressure relationship:

$$\rho = \rho_{Atm}\left(\frac{P}{P_{Atm}}\right) \tag{5}$$

where P_{atm} is the vapor density at the reference pressure P_{atm}. The resulting expression for radial flow to the vapor extraction well is:

$$\left(\frac{\varepsilon\mu}{kP_{Atm}}\right)\left(\frac{\partial P'}{\partial t}\right) = \left(\frac{1}{r}\right)\left(\frac{\partial}{\partial r}\right)\left[\frac{r\partial P'}{\partial r}\right] \tag{6}$$

where pressure P has been expressed in terms of the ambient pressure P_{atm} and the deviation from this pressure is P'.

The simplifying assumptions made prior to substitution of Darcy's law and the density-pressure relationship into the continuity equation include the following (Johnson et al., 1990):

● the flow system is isothermal;

● the vapors behave ideally;

● pressure changes do not significantly affect the porous medium structure relative to the porosity;

● the effect of gravity is negligible; and

- changes in the air-filled porosity due to contaminant removal are
  neglected.

Flow Under Vacuum Conditions

Under normal conditions, many factors influence the movement of gases
through the unsaturated zone. Diffusive transport of gases takes place in
response to a concentration gradient, according to Fick's law. Transport can
also occur due to density differences in pore gases; meteorological changes
such as changes in temperature, barometric pressure, and wind; infiltration of
rainfall; and a fluctuating water table.

All of these factors, however, are dominated by the pressure gradients
induced by a vacuum well in the vicinity of that well. Advective flow
resulting from the applied vacuum is far greater in magnitude than diffusive
flow, and thus, the above factors have a negligible effect on the flow to a
well during SVE.

The application of a vacuum to a well will cause a pressure gradient
(actually, negative pressure) to propagate throughout the zone in proximity to
that well. This zone extends radially away from the well for some distance;
this distance is known as the radius of influence. Many factors affect the
radius of influence. These include the strength of the applied vacuum; soil
properties like porosity and permeability; site features such as the
stratigraphy and the presence of an impermeable surface barrier or air inlet
wells; and other factors, which are discussed elsewhere.

Within the radius of influence, the vacuum is strongest at the well and
decreases as the distance from the well increases. Monitoring probes or wells
spaced throughout this zone allow the determination of the pressure field from
which isobars (contours of equal pressure) can be developed. Figure 7
illustrates airflow paths for an idealized extraction well application.

Flow occurs to the well in direct response to the vacuum. Flow lines
are usually perpendicular to the isopotentials; that is, flow occurs in the
direction of steepest pressure gradient. Pressure gradients are always
greatest near the well and least at a distance from the well. Therefore, a
surface seal might be of benefit in reducing air flow short circuiting near
the well (resulting in cleanup of a greater volume of soil) because it would
force air to be drawn from greater distances and to travel through more soil
to reach the extraction well.

AIR PERMEABILITY TEST METHODS

One critical factor used to determine the feasibility of SVE technology
is the vapor flow rate that can be induced at a particular site. The vapor
flow rate is directly dependent upon the air permeability (along with the
applied vacuum). Air permeability describes how easily vapors flow through
the soil. Since the air flow rate and the air permeability are linearly
dependent, a higher air permeability will result in a higher flow rate at the
same vacuum. Thus, there is a greater likelihood that SVE will be a feasible
remedial technology in soils with higher air permeability values.

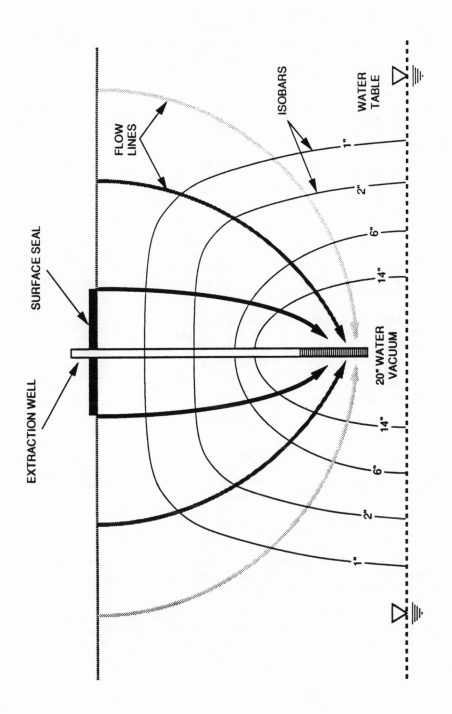

Figure 7. Relationship of Flow and Induced Vacuum

Field Methods

Numerous types of in situ methods have been employed for the determination of air permeability. All of these methods rely on measuring the difference between the ambient atmospheric pressure and the air pressure in the soil during vapor transport. Methods used include: air injection testing; oil field tests, such as pressure buildup and drawdown tests; and the procedure described by Johnson et al. (1990).

Air Injection. These methods rely on measuring the pressure differential created when air is injected into the soil. Equipment used for evaluation of surficial agricultural soils consists of a compressed air tank and a gas flow/pressure regulator attached to a cylinder that is inserted into the soil (Evans and Kirkham, 1949; Grover, 1955; Van Groenewoud, 1968). The pressure differential is measured by comparing the air pressure before and after a known volume of air is injected through the cylinder into the soil over a given time period. The soil air permeability is estimated using Darcy's law with both measured and known values for the pressure differential, air flow rate, cross-sectional area of the cylinder, and viscosity of the air. Air injection testing has also been used to test permeability in a packed-off section of a drill hole in granitic rock following a contained nuclear explosion (Boardman and Strove, 1966). The advantages of surficial air injection testing for air permeability are the portability of measurement equipment, rapid measurement, and low cost.

There are several limitations associated with using air injection testing to determine the air permeability for SVE tests. First, the permeability measured by injecting air into the soil may not be the same as that observed during implementation of soil vapor extraction. When air is injected into the soil, the soil particles will tend to expand and increase the apparent permeability. By contrast, the surficial soil may collapse during vacuum conditions experienced during SVE. Second, SVE requires air permeability of soils at depth, whereas air injection measures the permeability of the surficial soils.

Oil Field Test Methods. The petroleum industry has used several in situ test methods to determine soil air permeability. These tests and associated gas flow models are similar to the SVE vapor transport model presented by Johnson et al. (1990). Most of the tests are performed using one or more gas production wells. The natural gas from production wells is usually contained at higher pressures and temperatures than the gas subject to SVE. Thus, the modeling of natural gas flow and air permeability estimates of natural gas reservoirs usually incorporates parameters for gas compressibility and temperature not typically found in SVE vapor transport modeling. There are two commonly used methods of determining air permeability of gas producing formations: the pressure buildup test and the draw down test.

The *pressure buildup test* is conducted by closing a gas well that has been producing at a constant rate for a given period of time and monitoring the increase in down hole pressure after the well is closed (Donohue and Ertekin, 1982). The pressure buildup is considered to be the result of two superimposed effects: the pressure drawdown caused by the initial gas flow from the well and the increase in pressure that occurs when the well is closed. The pressure increase is modelled as a gas injection with the flow rate equal in magnitude but opposite in sign to the flow rate that was occurring during the well's active lifetime. The final pressure versus time relationship is approximated by the following equation (Donohue and Ertekin, 1982):

$$P_w{}^2 - P_r{}^2 - (1637\mu zTQ)/(km) \; \{log[(t + \Delta t)/\Delta t]\} \qquad (7)$$

where:

P_w - pressure in well (psia)
P_r - static reservoir pressure (psia)
μ - gas viscosity (cp)
z - compressibility factor (0<z<1)
T - reservoir temperature (°R)
Q - gas flow rate (MSCF/D)
k - permeability (darcies)
m - formation thickness (ft)
t - time at well shut-in
Δt - time interval over which pressure is measured

To find the permeability, $P_w{}^2$ is plotted versus $\{log[(t + \Delta t)/\Delta t]\}$ and the slope, A, equals:

$$A - -(1673\mu zTQ)/km \qquad (8)$$

The above equation is then solved for permeability, k.

The *pressure drawdown test* is another common oilfield method of determining permeability (Donohue and Ertekin, 1982; Ikoku, 1984; Smith, 1983). In this test, gas is extracted from a well at a constant flow rate while the pressure reduction in the well is observed over time. As above, the pressure versus time relationship is modelled using Darcy's law applied to radial flow. Additional factors are included to account for gas compressibility and "skin effects" (porosity variations near the well screen caused by well installation). The final pressure versus time relationship may be approximated by the following equation (Donohue and Ertekin, 1982):

$$P_w{}^2 - P_r{}^2 - (1637\mu zTQ)/km) \; \{log(t)+log(k/(\phi\mu cr^2)) - 3.23+0.87S] \qquad (9)$$

Additional parameters:

ϕ - porosity (fraction)
c - gas compressibility (1/psi)
r - radius of well (ft)
S - skin effect

For the drawdown test, the permeability is estimated by plotting $P_w{}^2$ versus log (t) such that the slope is the same as in equation 8. Permeability is determined from this equation.

The mechanics and modeling of the drawdown test are similar to those used by Johnson et al. (1990), described below. The gas compressibility and skin effects may be unnecesssary to determine the permeability during SVE air permeability tests. Pressure buildup and drawdown tests can be performed using a single well or, for increased accuracy, using both an extraction well and a monitoring well.

The soil air permeability *methodology proposed by Johnson et al. (1990)* for SVE technology can be considered as a special case of the drawdown test. In this case, the drawdown or vacuum pressure, P', is measured in a monitoring well at a distance, r, from the extraction well while a constant flow rate, Q, is extracted. The equation that approximates expected pressure changes over time, taken from Bear (1979), is:

$$P' = Q/(4\pi m)k/\mu) \quad \{-0.5772 - \ln[(r^2\epsilon\mu)/(4kP_{atm})] + \ln(t)\} \quad (10)$$

where:

   P' = vacuum measured in soil
   Q = extracted flow rate
   m = stratum thickness
   k = permeability
   μ = vapor viscosity
   r = distance to vapor monitor
   ϵ = vapor-filled void fraction
P_{atm} = atmospheric pressure
   t = time from start of test

As in the drawdown test, the permeability is measured from the slope of P' versus ln(t). Permeability is determined from the equation for the slope:

$$A = Q/4\pi m(k/\mu) \qquad (11)$$

To enhance the likelihood of obtaining meaningful soil air permeability data for SVE technology evaluation:

- take field measurements of vacuum pressure from monitoring wells around a vapor extraction well to allow for measurements at natural soil moisture content and to account for lateral heterogeneity of soil type and structure;

- maintain a constant air flow rate from the well; and

- conduct tests at several locations throughout the region to be remediated for increased confidence in the results.

Appendix E reprints the entire text of a paper by Johnson et al. (1989), which discusses this method in greater detail.

3. Site Investigations

INTRODUCTION

Federal underground storage tank regulations (40 CFR 280), promulgated on September 23, 1989, require that a site investigation be conducted when a release of product from an underground storage tank (UST) system has occurred or is suspected. The objective of the investigation is to characterize and delineate the area of soil and groundwater affected by the release.

The site investigation should address the following issues:

- The type of contaminants released (gasoline, fuel oil, solvents, etc.) and quantities released.

- The extent of product migration and routes for further migration.

- Product behavior in the subsurface environment (i.e., sorbed to soil, non-aqueous phase liquid (NAPL), dissolved in groundwater, or in vapor phase).

- Receptors subject to impact.

Johnson et al. (1989) have developed an investigation sequence that includes the following steps: (1) site history review; (2) preliminary site screening; (3) detailed site characterization; and (4) contaminant assessment. This generalized approach to investigation of leaking UST sites is one that regulatory staff and contractors should recognize as being similar to that of site assessment and remedial investigation approaches developed by EPA for uncontrolled hazardous waste sites.

Pursuant to the Comprehensive Environmental Response, Compensation, and Liability Act of 1980 (CERCLA) and the Superfund Amendments and Reauthorization Act of 1986 (SARA), technical assistance documents have been developed that provide a wealth of information on site investigation methodologies (EPA, 1987a; EPA, 1988c) and protocols (EPA, 1987b). These guidance documents address hazardous waste sites that may pose a more complicated investigation problem than a typical UST site due to the presence of numerous types of contaminants (caustics, metals, chlorinated solvents, PCBs, etc.) over a large area. Nevertheless, the investigatory approaches developed for these Superfund sites have application to leaking UST sites because they present a sound basis for data collection and assessment. This section provides a brief description of site investigation approach for leaking UST sites where implementation of SVE technology may occur.

41

SITE HISTORY REVIEW

The objective of the site history review is to assimilate available data that will be useful in guiding subsequent investigation activities. Data obtained should include historical records, site plans, engineering drawings, interviews with site personnel, meteorological data, boring logs, aerial photos, a soil survey and United States Geological Survey (USGS) maps. Record sources include the Town Hall, police and fire departments, EPA, state and local agencies, generator records and media reports. This information is useful in identifying the types of products stored in the USTs and the leakage and spill history of the site. Site plans and engineering drawings should be consulted to identify subsurface structures, pipes, and utility lines prior to any excavation, auguring and boring.

Soil survey maps, which can be obtained from the Soil Conservation Service (SCS), provide information regarding the types of soils at the site and environs. Many sites, however, have been altered and are composed of heterogeneous fill materials. USGS surficial and bedrock topographic quadrangle and geologic maps will indicate elevation contours and geology for the area. Geologic maps are useful for assessing potential migration routes.

The identification of potential receptors that may be impacted by the release is critical to any investigation. During the site history review phase, information on public and private water supplies in the area must be obtained. Locations of all municipal water supply wells should be noted on a site vicinity map to show their proximity to the UST site. The location of private water supply wells and surface water supplies should also be shown on a vicinity map.

Development of a generalized geologic cross section map will also be valuable during the initial investigation phases. The cross section should depict the on-site geologic strata and any nearby receptors such as public water supply wells. Even cross sections that lack detail provide a graphical presentation that helps to conceptualize potential migration routes and rates.

The site review should provide a basic understanding of the site to allow better use of time, equipment and laboratory analyses during the preliminary site screening and detailed site characterization phases. The site review should also aid in the assessment of health and safety procedures to be used by the field investigative team.

PRELIMINARY SITE SCREENING

The primary objective of the preliminary site survey is to identify the nature and extent of contamination. Preliminary site screening should include, at a minimum, a general site survey using field monitoring equipment, the development of a generalized site sketch and a subsurface environment cross section. The preliminary site screening may also entail a soil gas survey and groundwater sampling with mini-well points.

During conduct of the site survey the general location of the contaminated soils and the primary source of contamination should be noted as

well as the physical setting of the site.   During the site survey, potential
human health risks should be identified.   In addition, the general survey of
the site will provide some idea of the ground elevations of various
structures, locations of catch basins, location of possible sources of
contamination other than those initially identified, and soil properties such
as color and texture.   Ambient air volatile organic compounds (VOCs) should be
monitored to determine explosion risks and as an aid in locating the
contaminant source.   These surveys are conducted using field monitoring
equipment such as a flame ionization detector (FID) or photoionization
detectors (PID) (Spittler, 1980).   Gas chromatography (GC) can identify
specific compounds, but is often unnecessary at this preliminary stage.

The presence of catch basins on and off the site should be noted, as they
offer a convenient method for spilled fuel products to enter the soil zone.
Contaminants may also migrate along drainage culverts, ditches, sewer lines or
other conduits and then enter the soil zone through cracks or open joints.
Other potential sources of contamination, such as nearby service stations or a
user of hazardous compounds (such as a dry cleaning facility), should also be
identified during this stage.

Soil morphological features such as color and texture are useful
indicators of subsoil conditions.   Soils can be readily sampled to a depth of
six feet with a hand auger, and even greater depths (up to 20 feet) in sandy
soils.   Soil samples from various depths should be examined for color, texture
and morphological features, such as mottles.   Using visual assessment
techniques, the location of petroleum-stained soils, water table depth (if
shallow), and the presence of restricting layers or horizons in the soil
profile may be identified.

Determing the depth to water table is a key parameter in site screening
activities, since SVE is most effective at sites where the water table is 20
feet or greater in depth.   Where the water table is nearer the surface (less
than 5 to 10 feet), SVE may not be appropriate or, if used, may require some
means to lower the water table (Danko, 1989).   Gleying and mottling (variable
splotches of color in the soil) will indicate whether the water table is
permanent at that depth or fluctuating, respectively.

Values of hydraulic conductivity (Mishra et al., 1989) and residual
saturation of gasoline (Hoag and Marley, 1986), both important parameters in
system design, may be estimated from a particle size distribution analysis.
Preliminary field determination of soil texture can be used to make a crude
estimate of the permeability of the soil.   Generally, soils such as sands and
gravel are highly permeable and amenable to SVE.   Fine textured soils high in
silt and/or clay are more slowly permeable and SVE may or may not be
applicable.   The site investigator should note that contamination in the soil
profile at an UST site will probably include several different phases:

- Residual trapped in the porous media as the leaked product moved
  downward via gravity;

- Product volatilized into soil vapor;

- Free product floating on the surface of the water table; and

- Soluble components of contaminants in the groundwater or pore water.

Headspace analysis of soil samples is an additional means available for delineating subsurface contaminants. Soil samples may be collected with an auger and placed in wide mouth jars and sealed with aluminum foil. VOCs in the headspace of the sample can be determined by field monitoring equipment after the equilibration of the sample. Headspace analysis provides approximation of the concentrations of soil VOCs, however, and may not be highly correlated to soil contamination. Smith and Jensen (1987) found a poor correlation between field monitoring data and total petroleum hydrocarbon (TPH) concentrations in the soil. Even in view of these deficiencies, the approximate concentrations determined from headspace analyses will give some indication of the degree of contamination that must be addressed.

Soil gas is a screening tool that can be used to rapidly and cost effectively identify and delineate VOCs in the subsurface. A soil gas survey measures the VOC vapor phase concentration within the soil pore space. Figure 8 shows a typical soil gas sampling apparatus. Soil gas is most often used to optimize the placement of monitoring wells, to more precisely define an area designated for remedial action, or determine the most likely location of contaminant source areas (Thompson and Marrin, 1987) (Figure 9). Pitchford et al. (1988) used soil gas surveys at four sites to successfully map solvents, gasoline and JP-4 contamination. According to Kerfoot (1989), soil gas data is best used for planning the placement of SVE extraction wells.

The site specific usage of a soil gas survey should be based on factors such as soil type, depth to the water table and characteristics of contaminants as discussed by Marrin (1988). Coarse textured soils with low organic carbon content are typically more amenable to soil gas investigations. Fine textured soils (high in silt and/or clay) may retard vapor transport (diffusion and convection) due to their low air-filled porosity. Organic carbon in the soil will also retard vapor contaminant transport due to sorption of polar compounds onto the organic carbon. Soil gas is more applicable to sites where the depth to water is greater than 15 feet than to sites where the depth to the water table is shallower. This is due to a very steep chemical concentration gradient in areas with a shallow water table. In these areas, a slight variation in the ground elevation or depth to the water table can result in large variations in the measured concentrations of VOCs. Contaminants most applicable to soil gas surveys should have a vapor pressure of 1.0 mmHg or greater (20 degrees C). Additionally, contaminants should have an adequately low water solubility.

Soil gas measurements are limited in their capacity to predict soil and groundwater contamination. Bradford et al. (1989) found the predicative capabilities of soil gas concentrations to be valid only 1 to 8 feet from the soil or groundwater samples. Predictive capabilities also decreased with increased time between soil gas and soil or groundwater sampling. Karably and Babcock (1987) found that soil gas concentrations vary temporally and with climatic changes. Barriers in the soil profile, such as saturated clay

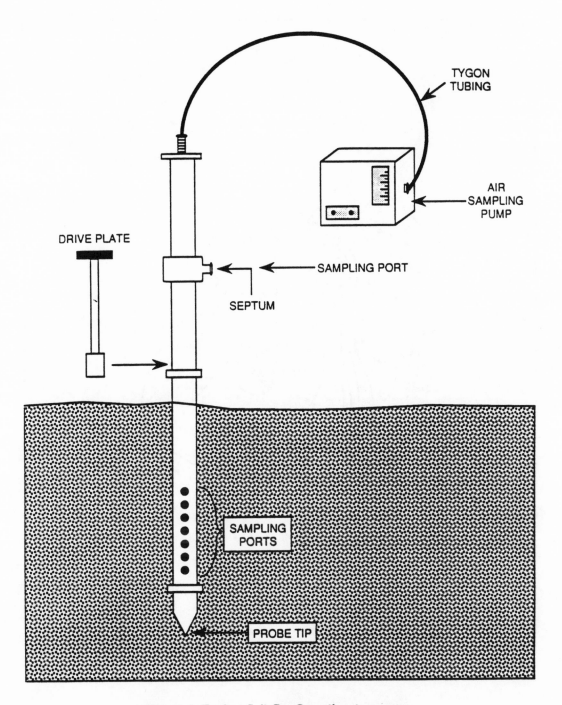

Figure 8. Typical Soil Gas Sampling Apparatus

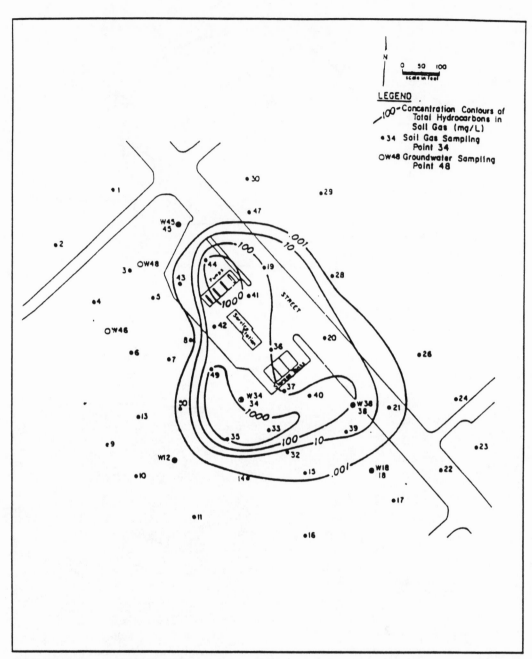

SOURCE: Thompson and Marrin, 1985

Figure 9. Soil Gas Total Hydrocarbons Concentration Contours

lenses, perched water bodies, pavement and buildings can affect soil gas concentrations (Kerfoot, 1989). Interpretation of results can be aided by preliminary modeling of diffusion transport (Silka, 1988).

Mini-well points allow for the in situ collection of groundwater samples prior to (or in place of) the use of monitoring wells. Micro-well points can also aid in the characterization of hydrogeologic conditions and assist with the optimum placement of permanent monitoring wells (Edge and Cordry, 1989). Well points can be quickly inserted into the soil and samples obtained for further analysis.

DETAILED SITE CHARACTERIZATION

If the preliminary site screening indicates that the product released is present in soil pore spaces and the product's volatility suggests that SVE technology may be an appropriate remedial action alternative, then additional site characterization geared towards SVE may be warranted. Methods for soil and groundwater sampling are outlined in the American Petroleum Institute's (API), "Manual of Sampling and Analytical Methods for Petroleum Hydrocarbons in Groundwater and Soil" (Kane, 1987).

Subsurface soil conditions may be determined from soil samples collected by an auger, split spoon sampler or Shelby tubes. Soil samples can provide information on soil texture and structure, density, and other diagnostic features. The SCS soil survey provides a detailed evaluation of the shallow soils, but only to a depth of six feet. Soil test pits provide a means to evaluate the horizonation of soils, to collect soil samples and to obtain soil cores for determining hydraulic conductivity in both the horizontal (K_h) and vertical (K_v) orientations. SVE is most effective in soils with hydraulic conductivity values above 1E-3 cm/sec. Contaminant removal, however, has been demonstrated in soils with hydraulic conductivities ranging from 1E-3 to 1E-6 cm/sec (Danko, 1989).

Moisture characteristic curves can be used to evaluate the characteristics of the soil under various soil matric potentials. Air-filled porosity can be determined from the moisture content and bulk density measurements. SVE is most effective in soils with a porosity between 40 to 50 percent (Danko, 1989). Additional soil tests to characterize the site with regard to SVE may include in situ hydraulic conductivity measurements and column studies. Hydraulic conductivity can be determined in situ using slug test techniques, pump tests, air entry parameters, or Guelph permeameters (Elrick et al., 1989). Soil cores may be used to evaluate the pore volumes required to remove specific contaminants as well as to determine air permeability on a microscale basis.

Geotechnical borings are generally necessary to identify deep strata and to allow sampling of soil and bedrock materials at depths of greater than five feet. Soil samples may be submitted to the laboratory for particle size analysis, bulk density testing, porosity determination, and chemical characterization. Geophysical techniques such as ground penetrating radar (GPR), electromagnetics (EM), resistivity, seismic methods, gravity, metal detection, and magnetometry can aid in the evaluation of subsurface conditions

(Benson, 1988). Site hydrogeology, including depth to water table, groundwater gradient and aquifer permeability, can be determined from monitoring wells and examination of the cores from split spoon sampling.

Monitoring wells are typically used to monitor VOCs in the groundwater along with other groundwater quality parameters. Groundwater is collected by either a bailer (stainless steel or teflon) or some type of pump (centrifugal, submersible, peristaltic or bladder). VOC samples are typically collected first, followed by collection of ABNs (semi-volatiles) and TPH (total petroleum hydrocarbons). Field determined parameters include temperature, redox potential, dissolved oxygen, pH, electrical conductivity and depth to water. Metals concentrations and general groundwater quality parameters such as total dissolved solids (TDS) and chloride are also readily collected.

CHARACTERIZATION OF CONTAMINANTS

The composition and concentration of contaminants in the soil, soil gas, and groundwater should be determined as part of the site investigation. These analyses will indicate the applicability of SVE as a remedial technology. Use of SVE for soil remediation is best suited to VOCs that exhibit a vapor pressure of at least 1 mmHg at 20 degrees C and a Henry's Law constant of at least 0.01 (dimensionless)(Danko, 1989). A survey of SVE use (Hutzler et al., 1989) has indicated that SVE is most commonly applied to gasoline and its volatile constituents (benzene, toluene, ethylbenzene, and xylenes), industrial solvents, including the dichloroethylenes, trichloroethylene (TCE), tetrachloroethylene (PCE), methyl ethyl ketone (MEK), and many others. Sites at which the contaminants are primarily high molecular weight petroleum fuels, such as fuel oil no. 6, or "weathered" gasoline are less amenable to SVE.

Monitoring often focuses on indicator compounds, such as BTEX (benzene, toluene, ethylbenzene, and xylenes), to narrow the scope of the analyses. Such indicator compounds should be easy to detect and be representative of the compounds at the site. Monitoring for certain indicator compounds is considerably easier than monitoring for the range of compounds that may be found at one site. Methods for various analyses are outlined in Kane (1987).

Alternately, a boiling point distribution can be measured for a representative sample. Compounds elute from a GC packed column in the order of increasing boiling point. A boiling point distribution curve can be constructed by grouping all unknowns that elute between two known peaks (e.g., between n-hexane and n-heptane) (Johnson et al., 1989) to determine the relative proportion of contaminant distribution.

Contaminants present in the soil should be characterized to determine applicability of SVE for remediation and overall contamination. Since some contaminants may not be removed by SVE, analysis of soils for less volatile contaminants (ABNs and TPH) may indicate the need for remediation by other methods (biotreatment, thermal desorption, soil washing or excavation). The cleanup goals must take into account the fact that only volatile compounds will be removed by SVE.

Characterization of contaminants in the soil gas will provide critical data regarding the contaminants present at the site. Often these data will provide clues about the release itself. By identifying the contaminants present in the soil gas the released product may be identified. This process may help to pinpoint the source of contamination - a leaking UST, a past spill at the site, or perhaps an off-site source. Identifying the source allows the contamination to be characterized more quickly and makes the site characterization more efficient. If an off-site source is identified, responsibility for clean-up could be assigned to a different party than was originally identified.

PILOT TESTING

If the site and contaminant characterizations indicate that SVE is an appropriate remedial measure, pilot testing may be performed to determine the site-specific design parameters for full scale design and implementation. The data obtained from the pilot test should enable determination of the radius of influence, initial and final exhaust concentrations, obtainable flow rates, water level changes, and vacuum well pressures. Collection of this information will be used to design the full scale system by determining: (1) number and location of extraction wells; (2) number and location of injection or inlet wells (if required); (3) equipment requirements, such as blower size (flow rate and vacuum); (4) aquifer parameters and groundwater pumping system specifications (if required); and (5) likely mass removal rates, and thus, the length of operation.

Data valuable for the design and operation of the SVE system should already have been collected during the site investigation. This data should indicate soil texture, organic carbon content, moisture content, bulk density, permeability and porosity from soil cores, groundwater elevations and the extent, concentration of the contamination in the soil, groundwater and soil gas. Soil textural data is needed to determine extraction and injection well design parameters such as filter pack media and well screen lot size. Textural data may also be used to estimate the hydraulic conductivity and residual saturation of the soil. Bulk density measurements can be used to estimate porosity. Permeability data can be used to estimate blower requirements and the radius of influence. The water table elevation relative to the contaminants should be used to determine the necessity of pumping equipment, based on an estimate of groundwater upwelling from SVE operation. The relative location of contaminants in the soil profile (shallow vs. deep) will determine the necessity for flow control devices such as surface seals and injection wells.

4. System Design

INTRODUCTION

A major advantage of soil vapor extraction technology is the relative simplicity of the design of these systems. In addition, the equipment that comprises the systems consists of commonly-used and widely available devices such as PVC piping, valves, and pumps. These factors impart an advantage to soil vapor extraction over other techniques (e.g., biotreatment or soil flushing) that may require more complex design or single-purpose equipment. Simplicity of design, however, does not imply that a logical, reasoned, and informed design procedure has been followed for all site specific installations. Maximum system efficiency and contaminant removal will occur only through a thorough understanding of the site and the SVE process.

The objective of a well-thought out and reasoned design process is to construct a soil vapor extraction system that removes the greatest degree of contamination from the site in the most efficient, timely, and cost-effective manner. The attainment of that objective will occur through an understanding of the three main determinants of system effectiveness (Johnson et al., 1989): the composition and characteristics of the contaminant; the vapor flow path and flow rate; and the location of the contamination with respect to the vapor flow paths. Design of an SVE system is basically a process to maximize the intersection of the vapor flow paths with the contaminated zone. Operation of the system should be done to maximize the efficiency of the contaminant removal and reduce costs.

This section discusses several aspects of SVE system design. Several options are available for layout of systems, including wells, trenches, and above ground soil piles. After selection of the appropriate system option, the number and placement of wells or trenches, the applied vacuum and pumping rate, the use of a surface seal or other types of air flow control, and the depth and size of the screened interval are all decisions that are made to maximize system effectiveness. Selection of the appropriate equipment type will also affect the system effectiveness. These aspects are discussed fully below.

EXTRACTION SYSTEM OPTIONS

Several options are available for extraction system layout. Figure 10 shows the most common methods: vertical wells, trenching or horizontal wells, and excavated soil piles. These variations are discussed separately below.

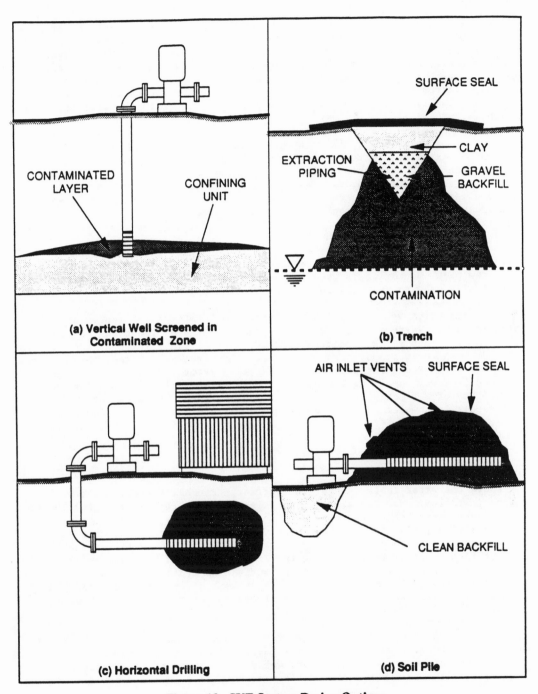

Figure 10. SVE System Design Options

Vertical Wells

Vertical wells are the most widely used SVE design method (Hutzler et al., 1989). This method is the only feasible option at sites where the contamination extends far below the land surface. Bennedsen (1987) suggests that horizontal wells or trenches may be more practical than vertical wells where the depth to groundwater is less than 12 feet. Vertical wells are generally inappropriate for sites with a shallow water table due to the potential upwelling of the water table that may occur after the application of a high vacuum.

Extraction wells are similar in construction to monitoring wells and, in many cases, existing monitoring wells have been used as extraction wells. Construction of an extraction well is straightforward (Figure 11). The bore hole is augured or drilled, PVC casing and screening (usually 2 inches to 12 inches diameter but depends on flow rate) are placed in the hole and the annular space is filled. Keech (1989) suggests that six-inch diameter extraction wells are more effective for removal of volatiles than the commonly-used two- or four-inch diameter wells. Slots are usually sized as small as possible to reduce silt entrainment. A highly permeable sand or gravel packing is placed around the screen for optimal gas flow to the well. Above the pack, bentonite is used to seal the hole. A cement-bentonite grout is typically used to seal the annular space to the surface.

The extraction well is typically located to intercept the center of contamination. Where multiple wells are used, they are placed so that the flow zone intercepts the contaminated zone. The screened interval should also coincide with the depth of highest product concentration. Often, this is just above the water table for products lighter than water like petroleum. The screened interval should be extended into the water table to allow for the possibility of a fluctuating water table. Also, the application of a vacuum will result in upwelling of the water table (Figure 12); if not counteracted, the wells may remove less vapor and more water, depending on the magnitude of the upwelling.

Trenches

Where the water table is near the surface, trenches or horizontal wells may be installed. Horizontal wells minimize the upwelling of the groundwater and allow coverage of a greater area than vertical wells. Installation of this type of well is accomplished quickly and easily where no surface or subsurface impendiments exist. A PVC drain pipe, wrapped in filter fabric to prevent fine material from clogging the drain, is placed at the base of the trench and backfilled with gravel. The surface is typically sealed with bentonite, asphalt, or a manmade liner to prevent air short-circuiting and infiltrating rainwater. The result is simply a dry trench or french drain (Figure 13).

Connor (1988) reports on the use of horizontal wells at a site where the depth to groundwater was less than six feet. DePaoli et al. (1989) used vertical wells, horizontal wells, and wells in an excavated soil pile at a jet-fuel spill site. Horizontal wells were installed in the excavated area

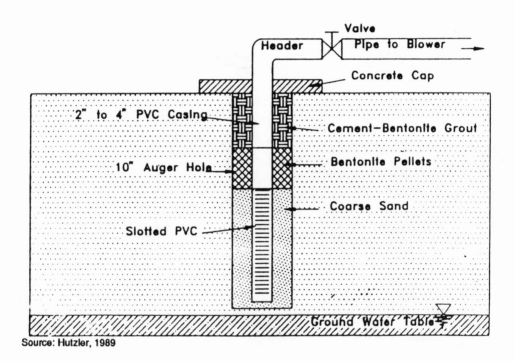

Source: Hutzler, 1989

Figure 11. Typical Extraction Well Construction

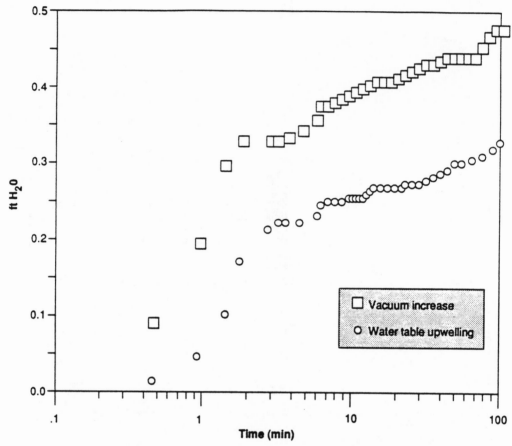

NOTE: (ft H$_2$0) denote vacuums expressed as equivalent water column heights

Source: Johnson et al.,1989

Figure 12. Water Table Upwelling in Response to Applied Vacuum

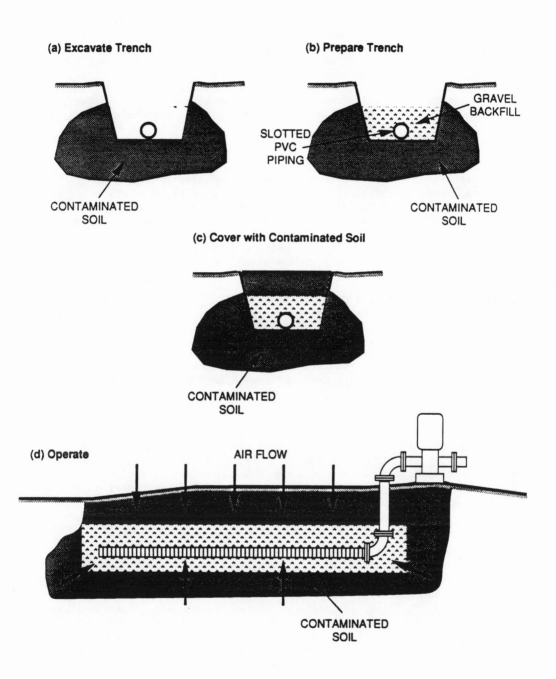

(a) Excavate Trench

(b) Prepare Trench

GRAVEL
BACKFILL

SLOTTED
PVC
PIPING

CONTAMINATED
SOIL

CONTAMINATED
SOIL

(c) Cover with Contaminated Soil

CONTAMINATED
SOIL

(d) Operate   AIR FLOW

CONTAMINATED
SOIL

Figure 13. Trench Construction

prior to backfilling to remediate any remaining contaminant. These wells have been operated only sparingly, however, so little data is available. Malot and Wood (1985) also report the use of horizontal wells. Horizontal wells can also be installed without excavation and backfill using special drilling techniques. Conventional drilling uses rigid drilling assemblies, whereas horizontal (lateral) drilling uses jointed, flexible drive pipe. While not widely practiced, there are companies such as Eastman Christensen (Houston, Texas) that specialize in horizontal drilling. Horizontal drilling possesses several advantages including those mentioned above, the ability to intercept specific zones, and the possiblility of intercepting more vertical fractures.

Excavated Soil Pile

Soil vapor can also be extracted from above-ground piles of excavated soil. This option is often used to remediate the soil removed from the leaking tank area at UST sites. The usual procedure is to excavate the contaminated soil and place it in a pile over one or more PVC pipes, which are packed in gravel and encased in filter fabric. An impermeable liner may be used to cover the excavated soil to prevent contaminants from volatilizing to the ambient air in an uncontrolled fashion and to prevent infiltration of precipitation. Prior to the system operation the cover is removed to allow air to be drawn into the pile.

Brown and Harper (1989) describe the use of vapor extraction in conjunction with biotreatment for excavated soil piles. A single, slotted PVC pipe was placed through the center of a six to eight foot high mound. DePaoli et al. (1989) report treatment of 52,000 ft^3 (1,500 m^3) of soil excavated during tank removal operations. The pile constructed was 160 feet in length and had a triangular shape, 43 feet wide at the bottom and 12 feet in height. At this site, eight vents were used, spaced 18 feet apart at a height of 5 feet.

WELL CONFIGURATION

Two main issues must be addressed with regard to the configuration of the extraction well(s). First, the number of wells required and their proper spacing and placement must be determined. Second, the extraction vents need to be sized and placed for optimal removal. Each of these topics is considered below.

Spacing and Placement

The number and locations of extraction wells required at a remediation site is highly site-specific and depends on many factors, including the extent of the zone of contamination, the physicochemical properties of the contaminants, the soil type and characteristics (especially the air permeability of the soil), the depth of contamination, and discontinuities in the subsurface. The radius of influence is the primary design variable and incorporates many of the above parameters. The radius of influence is the zone in which the effect of the vacuum is felt. Keech (1989) states that a vacuum of 0.1 inch of water or more in a monitoring well indicates that an extraction well has an influence at that point.

The initial step in placing extraction wells is to determine the radius of influence of one well. This is done by performing a field air permeability test, as described by Marley et al. (1989). The air permeability is the fundamental design parameter (Baehr et al., 1989), and is required to predict the effective area influenced by that well.

The objective of the test is to determine the pressure distribution throughout the vadose zone. To do this, one vapor recovery well should be located in an area likely to be remediated; this well can later be used during the actual cleanup. An existing monitoring well is often chosen. This recovery well should be surrounded by vapor monitoring wells or, more simply, soil probes, located at various distances and directions from the recovery well, and at varying depths. The object is to measure the vacuum induced at each location and compare that pressure to the flow rate at the recovery well. A simple method to do this is to start at the maximum test flow rate and decrease the flow slightly, stepwise, taking pressure measurements at the soil probes or monitoring wells at each step. Alternatively, the flow can be maintained at a constant rate and the vacuum measured against time. The results of this air permeability test are then plotted as shown in Figure 14.

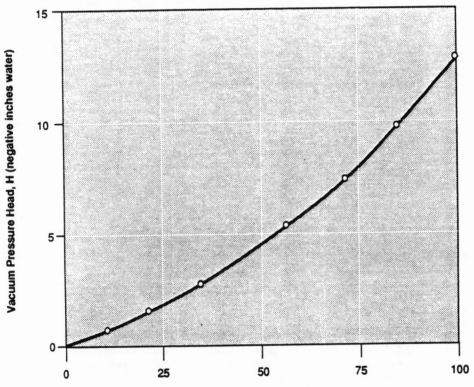

Figure 14. Field Air Permeability Test Results

The slope of the line would be directly proportional to the air permeability for one-dimensional flow conditions. For most SVE applications, however, radial flow conditions predominate and the determination of the permeability is more complex. These results can then be used to determine the effective radius of influence for each well. This is best done using a steady-state flow model and calibrating it against measured values from the field area permeability test. The results of the air permeability tests will allow a preliminary estimation of the radius of influence of one well.

The radius of influence of a well varies widely from site to site; Hutzler et al. (1989) report survey results that show the typical radius of influence ranges from 15 feet to over 100 feet. Generally, sandy soils result in smaller radii of influence than do clayey soils and require more closely-spaced wells at a given flow rate. However, the vacuum or flow may be increased for sandy soils. Johnson and Sterrett (1988) suggest that the radius of influence decreases as the bulk density increases and the soil porosity decreases. Krishnayya et al. (1988) modeled the effects of the suction head, the depth of the vadose zone, and the presence of a surface seal on the radius of influence. Results showed that the radius of influence increased linearly with increasing suction head. The depth of the vadose zone was also shown to highly affect the radius of influence; the radius increased non-linearly for increasing vadose zone depths. An impermeable surface seal was also modeled to determine the effects on the vapor flow paths. Results show that the presence of an impermeable surface seal increases the radius of influence, forcing air to travel longer horizontal distances and contact more soil than it would otherwise. Wilson et al. (1989) indicate that, as a rule of thumb, extraction wells should be spaced at two times the depth to which they are installed (e.g., if wells are 40 feet deep they should be spaced 80 feet on centers). This is highly site-specific, however. The location of the screened interval also affects the radius of influence and the vapor flow paths.

Extraction Vents

Extraction vents may be screened, slotted, or gravel packs. In addition to the vent construction, two other design decisions must be considered: the length of the vent and its location with respect to the unsaturated zone.

A goal of proper well design is to induce the air to flow through the zone of contamination to maximize cleanup efficiency. This is controlled by both well spacing and layout and by vent location. SVE operators use widely varying approaches to vent design, ranging from screening the entire depth from near the ground surface to the water table, to having a short interval at a particular depth corresponding to the zone of contamination. The location and length of screening will depend upon the stratification of the soil and the distribution of the contaminants in the soil. In many cases, the greatest concentration of petroleum vapors is immediately above the water table, especially at sites with a free product lens on the water table. Determination of the concentration gradient throughout the vadose zone can be accomplished with soil gas survey techniques.

In cases where the contaminant concentrations are greatest at the water table, the vents should be located close to or into the water table for optimal removal efficiency. In areas where the water table is expected to fluctuate throughout the period of venting, the screen length can be increased to ensure that venting can continue during periods of high water table.

The extraction wells are normally grouted or sealed with bentonite in the annular space to prevent ambient air from entering directly along the borehole/well interface. In some cases, the annular space is grouted down to the screened interval.

AIR FLOW CONTROL

In addition to the placement of the vacuum extraction wells, several other methods are available to control the flow paths of the extracted vapor to result in more efficient contaminant removal. These methods include the use of air injection or passive inlet wells, impermeable surface seals, and groundwater depression pumping. Each of these topics is discussed below.

Air Injection Systems

To enhance air movement through the soil, inlet or injection wells can be placed in strategic locations. Inlet wells are open to the atmosphere and allow air to be drawn passively into the soil from the surface. Injection wells use forced air to control the movement of air through the soil. Figure 15 shows the influence of inlet/injection well on the movement of air through soil.

Inlet or injection wells provide several advantages to SVE system design versus systems without air inlet wells. A major advantage is the ability to control the air pathways and thus, the zones of the soil to be affected by an SVE system. Connor (1989) stated that inlet wells act as a type of "vapor barrier"; that is, the radius of influence does not propagate beyond an inlet well in that vicinity. This allows the SVE system to be designed to give intensive treatment of a specific small area, rather than less intensive treatment of a larger area. The use of inlet or injection wells may also allow more rapid cleanup - by allowing greater flow rates - than would otherwise be possible. Injection of the extracted air may eliminate the need to obtain an air discharge permit. Also, if air is injected below the water table, volatiles may be "stripped" or volatilized from the dissolved phase into the soil gas. Disadvantages of their use include the added cost associated with construction of additional wells (although this would not be the case if inlet wells were just inactive extraction wells) and the added energy cost of the compressor for injection wells.

Connor (1988) reports the use of passive inlet trenches to increase the air flow rate to the recovery trenches. Gasoline recovery increased by a factor of three due to the improved pathway for air to enter the soil (ambient air had been restricted from entering the system through the use of an impermeable cap). Zenobia et al. (1987) describe a system with four wells that can be used alternately as extraction, injection, or inlet wells as the situation dictates.

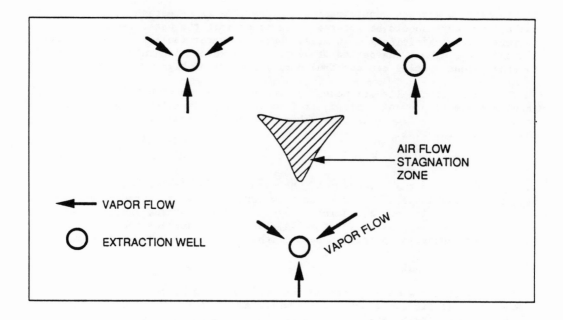

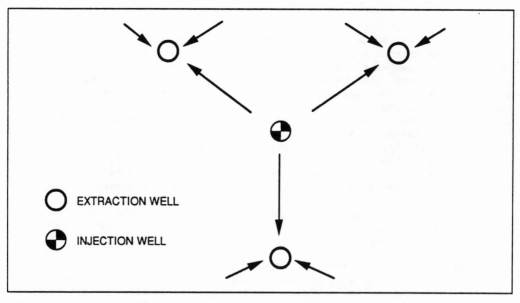

Source: Johnson et al, 1989

Figure 15. Effects of Air Injection on Flowpaths

Typically, injection and inlet wells are similar in construction to extraction wells, although injection wells may have a longer screened interval to provide uniform air flow. In fact, a well designed SVE system allows wells to act as extraction, injection, and/or inlet wells depending on the system requirements. Injection wells should be placed so that contamination is not forced away from the extraction wells in a manner that will result in bypassing the vapor treatment system.

Often, neither inlet nor injection wells are used, and air is drawn directly through the surface soils (Malot and Wood, 1985; Camp Dresser & McKee Inc., 1988). Using ambient air drawn through the soils is easy and inexpensive, but the air flow pathways are not readily controlled. Short-circuiting along pathways of product movement may occur, which would decrease the degree of contact between the contaminated zone and the fresh air. Short-circuiting refers to air being drawn down around the well without passing through smaller pores of the contaminated zone. In these cases, the use of impermeable barriers or surface seals may help to control vapor flow paths.

Surface Seals

An impermeable seal may be used where minimization of inflow from the surface is required (Figure 16). An impermeable surface seal prevents air from entering from near the extraction well (where the pressure gradient is the greatest) and forces air to be drawn from a greater distance and, ultimately, to contact a greater volume of soil. Surface seals may also prevent infiltration of rainfall, reducing the amount of water removed by the extraction well, thereby minimizing the production of air-water separator sidestreams. Surface seals also reduce fugitive VOC emissions from the soil to the air. Johnson et al. (1989) state that the effects of a surface seal are reduced when the screened interval is greater than 25 feet below the land surface. Krishnayya et al. (1988) modeled the effect of a surface seal and determined that the radius of influence increased dramatically but that the flow velocities decreased due to the lower pressure gradients (at a constant applied vacuum).

Depending on the characteristics of the site, different materials can be used as an impermeable cap. A flexible membrane lining can be rolled out on the site and can easily be removed when the treatment is complete. These membranes are available in a variety of materials, with high density polyethylene (HDPE) being the most common. These membranes are often used in landfill applications and are available from several companies. Selection of a synthetic liner that does not require backfill on top of the liner and is not susceptible to degradation by ultraviolet light will be most successful. For commercial or industrial sites, this option may not be feasible due to ongoing operations.

An alternative to a synthetic membrane is a clay or bentonite layer (Anastos et al., 1985). These natural liners can be applied to varying thicknesses. The drawbacks to clay liners are that they are not as easily removed as are the synthetic liners and they are more susceptible to damage from personnel and equipment. A third alternative, the most common at

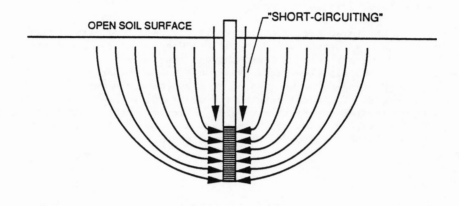

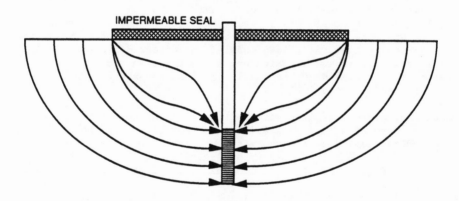

Figure 16. Effect of Surface Seal on Vapor Flowpaths

commercial or industrial sites, is the use of concrete or asphalt as a cap. This alternative is normally used only at a site that is already paved (Camp Dresser & McKee, Inc., 1988) or will be paved (e.g., a filling station).

Groundwater Depression Pumping

The vacuum induced in the vadose zone by the extraction wells will cause upwelling of the water table in the vicinity of the extraction wells about equal in magnitude to the vacuum at that point (expressed as inches of water). This may result in the groundwater being entrained in the vapor and being drawn up by the vacuum pump. This water must then be removed from the vapor stream via an air-water separator prior to passing the vapor stream through the pump or vapor treatment device. To prevent the entrainment of groundwater, Kemblowski (1989) suggests that a maximum vacuum of 20 inches should be used to minimize upwelling. Alternately, pumping wells may be used to depress the water table in the vicinity of the extraction wells (Figure 17). Pumping wells serve a dual purpose in cases where contamination exists near the water table. The depression of the water table will directly expose to air flow those contaminants in the capillary fringe and just below the original water table, allowing for volatilization of these contaminants. This method of operation makes SVE an especially effective complement to groundwater pump and treat schemes.

EQUIPMENT

The basic equipment for SVE systems consists of pumps or blowers to provide the motive force for the applied vacuum; the piping, valves, and instrumentation to transmit the air from the wells through the system and to measure the contaminant concentration and total air flow; vapor pretreatment to remove soil and water from the vapor stream; and an emission control unit to concentrate or destroy the vapor phase contaminants. Figure 18 shows a schematic diagram of a typical SVE system. This equipment is discussed below.

Piping/Blowers

The driving force for the creation of a vacuum in the soil is a positive displacement blower, a centrifugal blower, or a vacuum pump. Data presented by Hutzler et al. (1989) indicate that centrifugal blowers are about twice as common as vacuum pumps in SVE applications. Each method has particular advantages and disadvantages that should be considered at each site.

Several factors influence blower choice and design. Most important is the vacuum necessary to remove the design flow rate required to attain site cleanup within the time frame agreed upon. This depends on the areal extent of contamination and the air permeability of the soil. Pressure losses through the pump, piping, and the collection system in general will also affect the blower design. Many, widely varying values for the level of vacuum have been published, from just a few to over 100 inches of water (0.2 to 10 inches Hg). Weber (undated) gives general guidelines for a vacuum of 2 inches to 4 inches Hg for gravelly and sandy soils, 3 inches to 8 inches Hg for thick topsoil, and up to 14 inches Hg for clayey soil to produce an equivalent air flow rate. Vendors of blowers and pumps should be consulted for assistance to

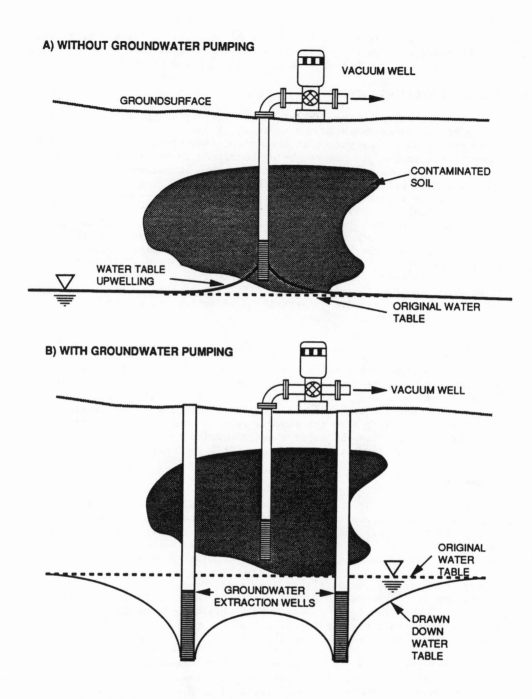

Figure 17. Groundwater Depression Pumping to Control Upwelling

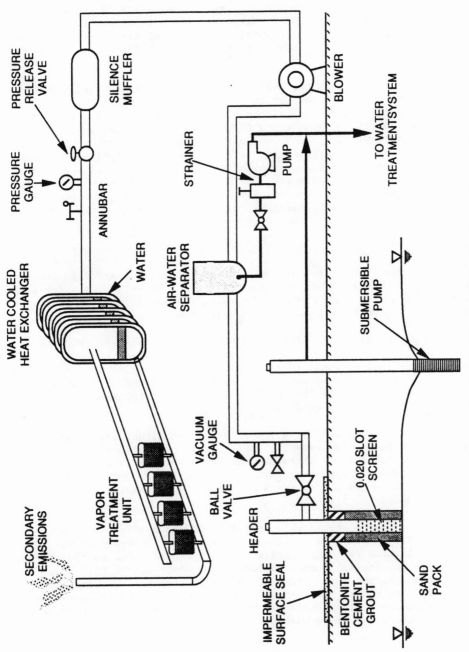

Figure18. Typical Soil Vapor Extraction System Schematic

help assess the size and number of pumps or blowers required.

Safety should be a prime consideration at any site where explosive gases are present. Motors that prevent or minimize vapor leaks should always be employed. If a vacuum pump is employed, the reduced pressure causes a reduction in the flash point temperature, so extra care is needed. Liquid-sealed vacuum pumps prevent metal-metal contact and reduce spark potential, further reducing explosion potential, but produce a wastewater stream.

The influent header from the wells to the pump should be equipped with a vacuum indicator, a manual flow control valve, and an ambient air bleed valve. The manual valve is used to control the blower and off-gas temperature to prevent overheating of the blower and discharge piping. Each individual well should have shutoff valves, vacuum indicators, sampling parts, and may be equipped with an air flow measurement device. Each type of blower and pump has its own temperature rise characteristics, and the choice of a blower with low temperature rise is often best. The effluent pipeline should be equipped with a pressure indicator, temperature indicator and automatic discharge valve. Hutzler et al. (1989) describe the type and rating of the pumps and blowers used at several SVE installations.

Piping

The piping used to connect the wells to the blower and emission control device is termed the manifold. Manifold piping may be very simple for a single well SVE system, but becomes increasingly complex for systems employing several extraction wells or systems with injection wells.

Manifold piping is constructed of either polyvinylchloride (PVC) (Schedule 40 or Schedule 80), polypropylene, HDPE, or stainless steel. The pipe diameter depends on the amount of total flow from all wells; six-inch piping is common. The piping can be above-ground or buried, which is common at active service stations. In the northern states, the piping may be insulated to prevent freezing of condensed water.

Proper piping design will have the riser sloped either back toward the well or towards liquid traps and the manifold piping sloped to liquid traps or sumps at various intervals. Entrained water and condensate (especially in cases where the above ground temperature is below the ground temperature), must be removed from the liquid traps occasionally, depending on the condensate produced.

The manifold system should also contain flow and pressure meters, to allow measurement during system operation, and flow control valves. Valves can be used to control which wells are active at any one time. One method of "pulsed venting" is to operate wells only periodically. This allows for vapor to diffuse into the larger pore areas. This may be accomplished simply by switching valve settings. Valves may also be used to bleed in ambient air to reduce vapor concentrations to comply with air discharge regulations or to maintain concentrations below unsafe levels.

Vapor Pretreatment

Vapors exiting the extraction wells may contain moisture and fine silt particles that may impair mechanical devices and vapor treatment operations. Air/water separators (knock-out drums and condensers), which often use demisting fabric and centrifugal force, are used to reduce moisture entering the vacuum pump and vapor treatment unit.

Knock-out drums (Figure 19) decrease the velocity of the incoming vapor stream, allow gravity to separate dirt and the heavier liquid molecules from the lighter vapor stream and change the flow direction to impinge particles on the side of the tank. Condensers use the same technology, and in addition refrigerate the vapor stream to condense any moisture in the vapor. Condensation may remove more contaminants than knock-out tanks if the concentrations of volatiles are above 5000 ppmv, but the added cost is often not justified.

Experience has shown that of the two components removed during vapor pretreatment - water vapor and water liquid - liquid water is by far more important. Thermodynamic calculations show that the amount of water vapor condensed from an air stream, assuming 98% relative humidity and a temperature reduction of 20 degrees C, will be approximately 0.1 gal/day/scfm. For a 100 scfm SVE system, the upper limit would be roughly 10 gal/day of condensate.

Liquid water entrained in the vapor during extraction may far exceed this amount. Malot (1989) reports liquid recovery of 1,500 gal/hr at a flow rate of 500 scfm. This equals 72 gal/day/scfm, almost three orders of magnitude higher than the theoretical maximum amount of condensate. The amount of liquid removed is highly variable and depends most directly on the proximity of the screened interval to the water table and the amount of infiltration. Application of a vacuum from the extraction well causes the water table to rise, especially in soils with high clay content; if the capillary fringe or water table rises enough to intersect the screen, significant quantities of water can be removed. Other important factors are the water content and porosity of the soil, extraction rate employed, and the size of the well.

Air/water separators extend the life of the vapor extraction system. Removing the moisture from the vapor stream will prevent the pump parts from rusting quickly. One vendor (Greene, 1990) suggests inclusion of a separator whenever the entrained water exceeds three gallons per minute (gpm). If activated carbon is being used as a vapor treatment process, reducing the moisture content is essential since high humidity will decrease the carbon's capacity to remove contaminants, vastly increasing carbon replacement costs. Piping and other systems are also susceptible to damage from extreme temperatures, although they can be protected by wrapping them with insulation. Air-water separators also provide an important barrier prior to the pump to prevent sediment and gravel from entering the blower machinery. This is especially important for blowers that operate at close tolerances.

A disadvantage of separating the moisture from vapor stream is that the side stream water must be treated. Many sites employ groundwater pump and treat schemes, such as air strippers, along with soil vapor extraction since

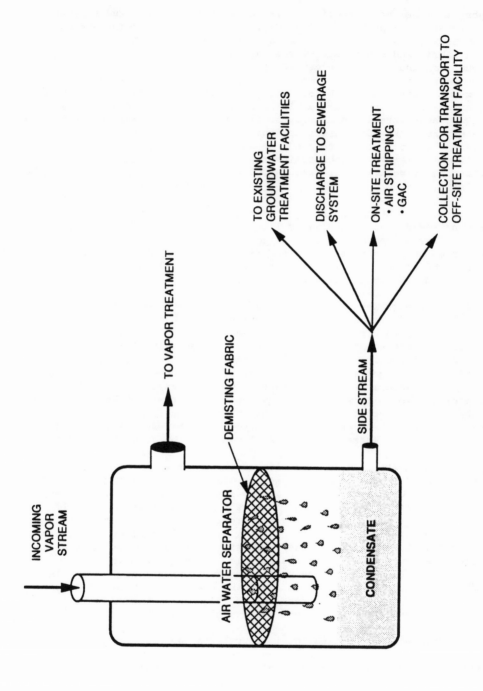

Figure 19. Vapor Pretreatment Sidestreams

the two methods complement one another. In such cases, the side stream water can be treated on-site by the facilities used for the groundwater treatment. An advantage of not separating the incoming stream is that this entrained water helps to cool the internal blower temperature, reducing the temperature rise through the blower or pump.

Discharge to a sewer system is another possibility for disposal of the side stream water. This option is viable only if there is a sewer line nearby, and if the concentrations of contaminants in the side stream water comply with local discharge regulations and treatment plant capacity limitations. The quality of this sidestream is poorly defined, however, with no references found in the literature. Permits must typically be obtained for this option. When neither of these two disposal options is available, the side stream water must be treated on-site separately or transported off-site by tanker truck for proper treatment and disposal.

Emission Control

Contaminants removed from the subsurface through SVE are not destroyed but rather only transferred to a different location. Many states and localities realize that without some type of treatment of the off-gas, soil vapor extraction may be simply substituting an air quality problem for a soil and water quality problem. For this reason, regulations regarding the disposition of the vapor phase contaminants are becoming increasingly widespread. The regulatory climate in the states and on the federal level regarding air discharge from SVE operations is discussed in the Introduction to this report. In general, treatment is now or soon may be required for all vapor streams except those that discharge minor amounts of contaminants into the atmosphere. Section 5 discusses in greater detail the most common options for treating the vapor phase contaminants.

Pre-assembled Systems

Some vendors now sell preassembled systems that can be hooked up to a well or manifold piping and incorporate all the above equipment. Often these units are trailer- or skid-mounted and can be brought directly to the site. Such systems have air/water separation, emission control, and may be operated by computer. These systems can usually be rented, leased or purchased outright.

5. System Operation and Monitoring

INTRODUCTION

Once installed, SVE systems can be operated in a flexible manner for optimal removal efficiencies. The systems should be operated to optimize both efficiency and effectiveness so that the greatest amount of contamination is removed in the shortest time period at the lowest cost. Monitoring systems, which are essential for determining SVE system success and cleanup attainment, are composed of equipment familiar to most operators. Interpretation of the monitoring data, however, is sometimes a complicated task. The clean up attainment determination depends on having set cleanup goals prior to system startup and proper monitoring procedures and interpretation of those data. This section provides a brief overview of SVE system operation and monitoring considerations.

SYSTEM OPERATIONS

Once the SVE system is designed, constructed, and installed, the start-up consists of turning on the SVE blower(s) or vacuum pumps and, as appropriate, opening the system air inlet wells to the atmosphere. Vacuum gauges installed at various locations on the wells and manifold network are monitored during start up so that flows and pressures can be adjusted to be compatible with the system design. Several hours to several days of system operation are required to establish steady-state flow conditions, depending on the air permeability of the formation (Johnson et al., 1990). After the start-up period, the SVE system may be left in continuous operation essentially unattended except for daily checks to make sure the water level in the air-water separator does not rise above the safe level and occasional tank draining. In addition, the blower must be serviced periodically by checking the drive belts and lubricating the bearings. The emission control unit may require more extensive attention, especially during the early stages of operation. In general, maintenance requirements are highest at system startup and decline over time.

The VOC extraction rate is measured by sampling VOC concentrations in the exhaust air and measuring flow. Removal rates, typically expressed in units of mass per time, equal the concentration multiplied by the volumetric flow rate. Numerous researchers have shown that the rate of VOC extraction is high initially, but decreases with time. This decrease may signal the transfer to a diffusion limited system. In other words, the saturated vapors present in the soil at system startup are quickly removed. Removal of contaminants thereafter may be diffusion-limited. Since diffusion rates are slower than advection, removal rates drop with time of continuous operation.

To maximize VOC removal while minimizing pumping, the vacuum pump may be shut off and the soil vapor allowed to re-equilibrate. This method is known as "pulsed venting". Alternately, different combinations of wells may be vented to change the flow field and accomplish the same goal. VOC concentrations increase after a temporary shutdown, although to a level lower than the concentration at the initiation of vapor extraction. Figure 20 shows an example of data collected during pulsed venting.

Several studies have indicated that intermittent venting from individual vents is more efficient than continuous operation in terms of mass of VOC extracted per unit of energy expended (Hutzler et al., 1989; Crow et al., 1987; Oster and Wenck, 1988; Payne and Lisiecki, 1988). Optimal operation of an SVE system may involve taking individual vents in and out of service to allow time for liquid and gas diffusion and to change air flow patterns in the region being vented (Hutzler et al., 1989). For example, blowers can be equipped so that they could be moved easily from one well to another for a cost effective means of operation.

ENHANCED BIOTREATMENT

An additional benefit of vapor extraction systems is the enhancement of biodegradation of organic contaminants. The increased amount of oxygen in the soil pores that results from SVE operation seems to be responsible for the increased biological activity. Figure 21 shows the oxygen increase following SVE startup for one site (Hinchee et al., 1989). Biodegradation can be monitored by measuring the ratio of oxygen to carbon dioxide (O_2/CO_2) (Johnson, R.L., 1989). Connor (1988) showed that soil temperatures increased in the zone of contamination, presumably as a result of biodegradation. At a jet fuel spill clean-up in Utah (DePaoli et al., 1989), CO_2 and O_2 levels were measured in the soil gas. Initially, CO_2 levels were high (11 percent) and O_2 levels were low (1 percent). As venting continued, the CO_2 levels decreased and O_2 levels increased. Carbon dioxide levels continued to be an order of magnitude higher than background. Considering that the hydrocarbons are the only source of carbon at this site (the site was underlain by sandy soils) and that half of the hydrocarbon consumed by bioactivity is converted to CO_2 (the other half is converted to biomass), it was calculated that bioactivity contributed 27.5 percent of total hydrocarbon removal at this site (DePaoli et al., 1989). Similarly, Fall and Pickens (1989) report that at another site biodegradation accounted for approximately 40 percent of removal. Figure 22 shows how biodegradation contributed to the overall mass removal at a site in Utah. Hinchee (1989) suggests designing SVE systems to maximize degradation and minimize volatilization of contaminants to decrease vapor phase treatment costs.

SYSTEM MONITORING

SVE performance must be monitored to insure efficient operation and to determine when it is appropriate to shut off the system. Johnson et al. (1989) recommend that at a minimum, the following parameters should be measured and recorded:

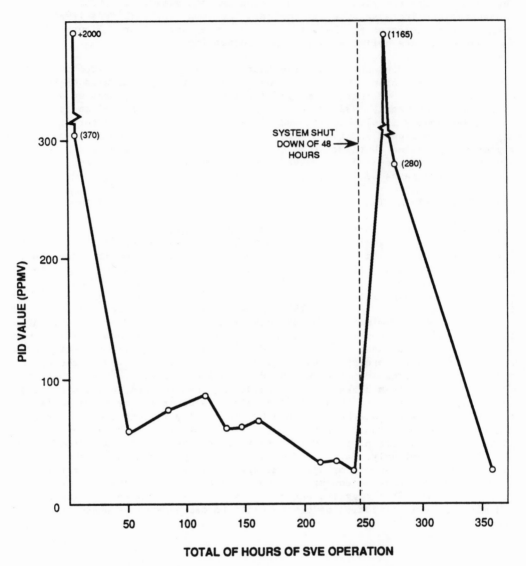

Source: Zenobia et al., 1987

Figure 20. Effect Of "Pulsed" Operation

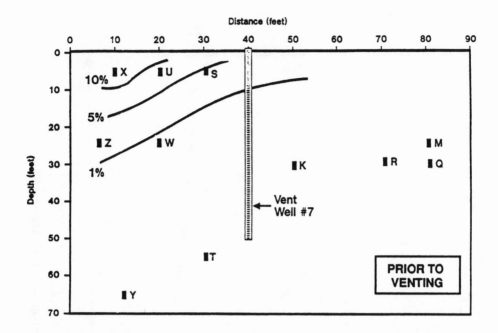

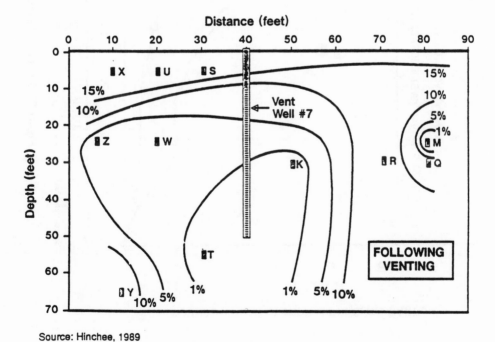

Source: Hinchee, 1989

Figure 21. Oxygen Increase Following SVE System Startup

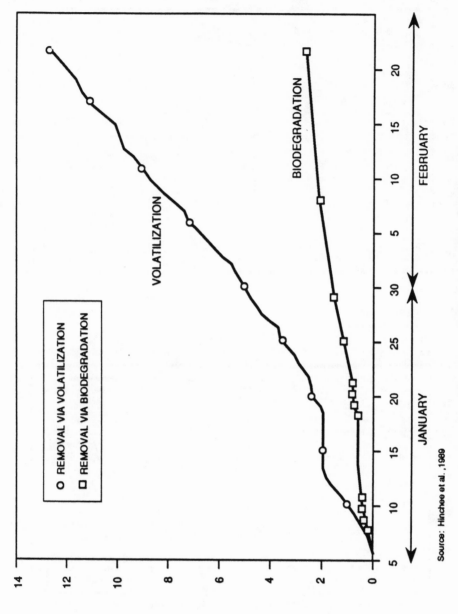

Figure 22. Effect of Biodegradation on Overall Removal Rate

●   Date and time of measurements.

●   Vapor flow rates at each extraction well and injection well.
      Measurements can be made by a variety of flow meters, including
      pitot tubes, orifice plates, and rotameters.

●   Pressure readings at each extraction and injection well can be
      measured with manometers and magnehelic gauges. Pressure should
      also be monitored at each soil gas probe location.

●   Vapor concentrations and composition from each extraction well.
      Vapor concentrations can be measured by an on-line total
      hydrocarbon analyzer calibrated to a specific hydrocarbon. This
      information can be combined with vapor flow rate data to calculate
      removal rates (mass/time) and the cumulative amount of contaminant
      removed. Soil gas measurements should be made periodically at
      different radial distances at soil gas probes to monitor the
      reduction in contaminant vapor concentration.

●   Temperature of the soil and ambient air. (Connor (1988) predicted
      through monitoring soil temperatures that biodegradation was
      occurring in the zone of contamination.)

●   Water table elevation (for soils with a relatively shallow water
      table). Water level measurements can be made with electronic
      sensors located in air tight monitoring wells.

●   Meteorological data, including barometric pressure, precipitation,
      and similar data.

In addition to these parameters, product thickness on the groundwater
should be measured if more than one-eighth inch exists on the water table.
Capillary rise in the soil can result in an apparent free product thickness 25
percent or more greater than the actual free product thickness (Hoag and
Cliff, 1985).

Regalbuto et al. (1988) observed that where the vadose zone is made up
of clean, coarse-grained material, most of the removal will typically occur
within the first three months of operation. This is due to the removal of the
most volatile fraction of the product early in the system operation. The
extraction rate was not necessarily highest immediately after SVE start-up.
During the first few months of operation the VOC concentrations varied as much
as an order of magnitude from one day to the next.

VAPOR STREAM CHARACTERISTICS

The physicochemical properties of the contaminant-laden vapors change as
the vapors move from the soil through the SVE system until they are ultimately
discharged to the atmosphere. This section discusses those properties and how
they vary through the system.

Extraction Wells/Manifold Piping

Subsurface soil vapors are usually saturated with water in the eastern
portion of the U.S. In the arid regions in the Southwestern United States,
the soil vapors may not be as moist, but relative humidities (RH) of 60
percent or more are common. This is particularly true when the water table is
relatively close to the area of the screen in the venting wells. If the
ambient air temperature is lower than the subsurface temperature (e.g., winter
conditions), the water in the soil vapor will begin to condense as the
extracted vapor reaches the surface piping of the SVE system, resulting in
free water droplets in the vapor stream. Piping from the extraction well may
be sloped back towards the extraction well at a shallow slope of approximately
one percent to induce flow of accumulated moisture back into the well.
Alternately, the condensate can be collected in liquid traps and then treated.
In vapor wells screened close to the water table that have high vapor
extraction velocities, water droplets may carry-over as a mist into the
manifold piping due to the mechanical energy of the vapor stream. This free
water in the stream must be removed because it may cause operational problems
with the blower, and reduce the effectiveness of air emissions control
processes such as carbon adsorption units.

Air/Water Separator

The free water in the air stream is usually removed with the use of an
air/water separator (Figure 19). This unit usually consists of a tank with a
demisting fabric in the inlet. The air/water separator in the line works by
decreasing the velocity of the vapor stream, thus allowing the free water and
sediment particles to drop out and collect in the tank. The demisting fabric
helps the suspended droplets collect and drip into the tank without being
carried over past the air/water separator. Although, the air leaving the
air/water separator is usually still saturated (relative humidity - 100%), the
free water in the line is removed using the separator. The air/water
separator also serves to collect sediment and gravel that may be removed from
the wells during pumping.

Vacuum Pump/Blower

The pump or blower causes a pressure differential between the inlet and
outlet streams that results in a vacuum in the subsurface. The heat of
compression in the blower causes an increase in temperature and a
corresponding decrease in relative humidity (although the total mass of water
remains constant) in the vapor stream as it exits the blower. This
temperature increase may cause operational and efficiency problems in GAC
systems if not reduced.

Heat Exchanger

The heat of compression of the vapor through the blower causes the vapor
stream to increase in temperature to a level which may reduce the efficiency
of the carbon adsorption units. This effect is often minimized by cooling the
vapor stream with a heat exchanger to a temperature that can achieve an
optimal condition between air temperature and level of saturation. The heat

exchanger can be designed to cool the vapor with cooling water, with forced air, or with another liquid such as brine or glycol. Heat exchangers are often used to preheat incoming vapors using the heat of exhausted vapors that have passed through thermal incineration or catalytic oxidation systems.

Emissions Control System

The emissions control system is the mechanism by which the contaminants in the vapor stream are treated prior to discharge to the atmosphere. Vapor combustion units use heat to combust flammable contaminants. Incineration is recommended for use on vapor streams containing high concentrations (>10,000 ppmv) of contaminants such as BTEX compounds. Destruction efficiencies are typically 95 percent or greater.

Catalytic oxidation units operate by passing the heated vapor stream over a catalyst bed which facilitates combustion. Destruction efficiencies are typically greater than 95 percent. When this method is used for vapor streams containing more than about 3000 ppmv (depending on the heat value of the influent stream) of volatiles, the vapor stream must be diluted with fresh air because more concentrated vapors can cause catalyst bed melt-down. Although best suited for petroleum hydrocarbon compounds, catalysts have been developed which can break down halogenated hydrocarbons as well. The technology for chlorinated hydrocarbons is relatively new, and further treatment of the emissions stream may be required with scrubbers to prevent releases of acidic by-products.

Activated carbon adsorption is a proven technology which has been used for many years for treatment of vapor streams containing organic compounds. The process works by passing the vapor stream through a bed of activated carbon. The carbon has a high specific surface area (ratio of area to volume), which allows a large surface to adsorb contaminant compounds. For optimal efficiency, the relative humidity should be below 50 percent. Contaminant saturated carbon is potentially hazardous and must be handled with care.

CLEAN-UP ATTAINMENT DETERMINATION

The determination that a site has been "cleaned up" involves both having a specific cleanup target and then actually measuring or monitoring for that target level in the field.

Target Levels

Generally, soil clean-up target levels are pre-determined on a site by site basis and depend on state and local regulations and guidelines. Confirmatory soil borings and, in some cases, soil gas samples are required prior to closure. Soil analysis is usually expensive and may be disruptive to the site; therefore it is judicious to determine where soil borings should be taken. Key criteria for determining when the system can be shut down include: cumulative amount of contaminant removed, extraction well vapor concentrations, extraction well vapor composition, soil gas contaminant concentration and composition (Johnson et al., 1989), or remaining soil

concentration. Appendix K contains the results of a survey of the states'
requirements regarding soil cleanup attainment. It appears that most states
limit the allowed residual that may remain in the soil (typically expressed as
TPH). Cleanup is considered complete when soil samples indicate that the
residual product is below the regulated limit. When setting clean-up
standards for SVE sites, it should be noted that immobile,
high-molecular-weight compounds will remain in the soil and will be identified
during soil analysis. These compounds are the least mobile compounds,
however, and pose the lowest exposure risks (via the groundwater pathway) of
all petroleum compounds.

Measuring the vapor concentrations in the extracted vapors gives an idea
of the efficacy of the system; however, a decrease in vapor concentrations is
not necessarily strong evidence that soil concentrations have decreased.
Decreases in vapor concentrations can also be attributed to other phenomena
such as water table upwelling, increased mass transfer resistance due to
drying, diffusion-limited mass transfer flow from dead-end zones, or leaks in
the extraction system ("short-circuiting") (Johnson et al., 1989). Monitoring
of the extraction well vapor composition as well as the concentration gives
more insight into the effectiveness of the system. If the total vapor
concentration decreases without a change in composition then it is most likely
due to one of the phenomena listed above. If the decrease in concentration is
accompanied by a shift in composition to less volatile compounds, then it is
most likely due to a change in the residual contaminant concentration.
DePaoli et al. (1989) show how the composition of the extracted soil gas
changed over the course of system operation (Figure 4). As the total volume
of gas extracted increased, the fraction of lighter-end components decreased
while the fraction of heavier components increased markedly. The figure shows
that the lighter, more volatile constituents (such as butane and propane) are
removed first. As more vapor is removed, the heavier, less volatile
constituents predominate due to the absence of the light fraction. This
phenomenon influences cleanup criteria in some states. For example, a residual
gasoline cleanup operation might operate a vapor extraction system until
benzene, toluene and xylenes were not detected in the vapors. The remaining
residual would then be composed of larger, less mobile molecules, which would
not volatilize or leach as easily and therefore may pose less of a health
threat (Johnson et al. 1989).

Residuals Measurement

The residual petroleum in the soil can be measured directly, by
retrieving a sample and then determining how much petroleum remains, or
indirectly, by monitoring the soil gas concentration and composition.. Often,
the soil gas measurements are made throughout the course of the cleanup, and
that data is used to determine where the final soil samples should be taken.

Residuals analysis typically means the quantification of either Total
Petroleum Hydrocarbons (TPH), total Volatile Organic Compounds (VOCs), or one
or more indicator compounds (e.g., benzene, toluene, ethylbenzene, and xylenes
(BTEX)). Most state and local regulations are based on these measurements
(Table 5). These parameters are determined in different ways, each of which
has its particular limitations.

TABLE 5. CLEANUP LEVELS PUBLISHED FOR PETROLEUM CONTAMINATED SOILS

State	TPH (ppm)	Notes
AK	NE	Clean to background or use LUFT to justify level.
CA	10-1,000	Leaching analysis potential - LUFT manual.
CT	NE	Level necessary to protect groundwater.
FL	5	Sum of BTEX <100 ppb.
HI	NE	Function of water quality standards.
IL	NE	Function of health considerations
KS	NE	Function of groundwater standards.
KY	NE	To background or detection limit.
ME	20-50	Level dependent on water quality factors.
MD	NE	Case-by-case basis.
MA	100	Or 10 ppm total organic volatiles as benzene.
MI	NE	To background levels.
MS	NE	Function of water quality standards.
NH	<10	BTEX <1 ppm for gasoline and diesel.
NJ		BTEX - Groundwater based standards
NY	NE	Function of water quality considerations.
PA	NE	Case-by-case basis.
RI	50-100	TPH level function of site specific factors.
SC	NE	Case-by-case basis, function of water quality.
TN	100	10 ppm total BTX in soil (except for gasoline.)
TX	NE	Function of water quality standards.
UT	NE	Standards under development.
VT	<10	May apply water quality standards.
WI	<10	Groundwater standards for benzene and toluene.
WY	<10,<100	Depends on depth to groundwater

NE = None established   *Adapted from Bell et al. 1989

The measurement of TPH begins with the extraction of oil and grease from the sample using Freon 113, which is the trade name for a fluorocarbon solvent. The oil and grease is composed of both fatty materials from animal and vegetable sources and hydrocarbons of petroleum origin. The Freon 113 is then passed through a polar silica gel, which adsorbs the fatty materials, leaving the hydrocarbons in the Freon 113 solvent (Kopp and McKee, 1979).

The hydrocarbons are then quantified by gravimetric techniques or infrared (IR) spectroscopy. The IR spectra of petroleum hydrocarbons contain a variety of bands that are indicative of specific molecular structures. The advantage of the IR method is that instrumentation costs are modest (Potter, 1989). Unfortunately, precision and accuracy vary greatly depending on the petroleum products undergoing analysis. For example, the more volatile components of gasoline or light fuel oils may be lost in the solvent concentration step. The recoveries for heavier distillates are often low because many of the distillate constituents are only sparingly soluble in Freon (Potter, 1989). Additionally, environmental concern over the use of chlorofluorocarbons is growing and the long-term availability of Freon is uncertain.

Quantification of volatile compounds such as BTEX is usually carried out using gas liquid chromatography or gas liquid chromatography coupled with mass spectrometry, which can detect specific organic cOompounds. Volatile compounds from the sample may be identified and quantified using "purge and trap" techniques. Limitations of this analytical method include the detection of non-target compounds and poor chromatographic resolution with packed and capillary columns (Potter, 1989). BTEX are typically used as indicator chemicals for assessing the extent of leaking gasoline USTs (CDM, 1987). However, if the dissolved gasoline plume in the ground is older than one to two years and free product no longer exists, the BTEX compounds may no longer be present because they are readily biodegraded. Under these conditions, other gasoline compounds such as methyl-tertiary butyl ether (MTBE), an octane enhancing additive found in some gasolines, could be used to locate and delineate the plume.

Soil humic materials can influence the results of both TPH and BTEX analytical tests. Many hydrocarbon compounds that are found in various petroleum products occur naturally in soil (Table 6). Aromatic hydrocarbons such as benzene, toluene and xylene occur naturally in soils at concentrations of 1-5 ppm (Dragun and Barkuch, 1989).

Given the dynamic nature of the soil's biota and chemical reactivity, the petroleum products themselves undergo alteration. As petroleum products weather in the soil environment they may not reach a mineralized endpoint. Instead, the degradation process may result in relatively stable aliphatic and aromatic compounds, which are integral parts of soil humus (Alexander, 1977). Because both humic and fulvic acids, fractions of humic material, are composed of random polymers of aromatic and aliphatic substituents (Tate, 1987), petroleum hydrocarbon degradation processes may increase the amount of soil humic material.

TABLE 6. PETROLEUM HYDROCARBON CONSTITUENTS THAT OCCUR NATURALLY IN SOILS

acetic acid	alkanes
benzene	1,2-benzofluorene
benzoic acid	butanoic acid
carbazole	decanoic acid
2,6-dimenthylundecane	n-docosane
n-dotriacontane	n-eicosane
eicosanoic acid	ethanol
ethylbenzene	formic acid
n-hemeicosane	n-hentriacontane
heptacosane	n-heptadecane
heptanoic acid	hexacosane
n-hexadecane	hexadecanoic acid
methane	methanethiol
methanol	naphthalene
n-nonacosane	n-nonadecane
nonanoic acid	n-octacosane
n-octadecane	octanoic acid
pentacosane	n-pentadecane
pentanoic acid	perylene
phenanthrene	propanoic acid
n-tetracosane	n-tetradecane
tetradecanoic acid	toluene
n-triacontane	n-tricosane
p-xylene	o-xylene

Source: Dragun and Barkach, 1989.

Given the potential for the enhancement of soil humic material via petroleum degradation, questions arise concerning what the determination of TPH is measuring. Is it measuring actual petroleum product hydrocarbon compound and/or the end products of biotic and abiotic transformations of petroleum products? There appears to be no published research describing compounds to which petroleum hydrocarbons may be transformed or methods to distinguish between humic material and petroleum hydrocarbons. New techniques and methods that could address some of these issues are discussed below.

The EP Toxicity Test is designed to simulate the leaching to which a waste would be subjected if disposed of at an improperly designed sanitary landfill. This procedure can be applied to solids, liquids, and multiphasic materials. The application of this test will identify wastes that have the potential to leach significant concentrations of hazardous compounds. Following the extraction procedure the leachate is analyzed. This method, since it is designed to extract only inorganics, is not applicable to hydrocarbon spill sites. This disadvantage has led to the development of an extraction method that is capable of leaching both inorganics and organics from a waste sample.

The Toxicity Characteristic Leaching Procedure (TCLP) was designed to improve the EP toxicity protocol and to expand its applicability to a greater number of contaminants that could leach from waste materials. This procedure is used to extract semi-volatile organic compounds, pesticides and metals using sample containers similar to those used in EP toxicity analysis. However, with the utilization of a "zero headspace extractor", volatile organic compounds can be extracted (EPA, 1986b). The key difference between the EP toxicity method and the TCLP method is that TCLP uses a more acidic leaching medium for moderate to highly alkaline wastes. This ensures that the leaching potential is not underestimated. The leaching medium is acetic acid in a sodium acetate buffer solution at a pH of 2.9 or 5.0. For moderately to highly alkaline wastes the 2.9 pH solution is used, while the 5.0 pH solution is used for other wastes. The extraction procedure is identical to EP Toxicity, except that the agitation step requires the use of a rotary end-over-end shaker, the extraction time is 18 hours, and the extraction procedure must employ a "zero-headspace extractor".

The limitations or sources of variability associated with any leaching procedure includes:

- chemical differences in the laboratory leaching media and the actual leaching solution at the site,

- liquid:solid ratio, and

- number and time of extraction(s).

The advantage of using a leaching test is that the question of identifying the substance is secondary and the issue of risk to the environment is directly addressed.

6. Secondary Emission Control

INTRODUCTION

SVE systems that do not incorporate a treatment system for the extracted vapors do little other than transfer the contaminant from the subsurface environment to the atmosphere. One may argue that the environmental risks associated with atmospheric discharge of the extracted vapors are less than those associated with leaving the vapors in the soil. Dispersal of these vapors in the atmosphere renders the concentrations harmless, this argument holds, while leaving the contaminants in the soil means a continued risk of migrating to water supplies. This argument is spurious, however, considering the air quality problems being encountered in some areas, the global impacts now being attributed to hydrocarbon and chlorinated organics releases, and the availability of effective vapor treatment technologies. Regulatory agencies are becoming increasingly stringent with regards to SVE air discharge and permitting requirements as the technology becomes more widely used.

This section discusses the six most common alternatives for dealing with the evacuated vapor stream from an SVE system:

- adsorption on granular activated carbon;

- thermal incineration;

- catalytic oxidation;

- internal combustion engine;

- biodegradation; and

- direct discharge.

Many of these methods have been used in industrial applications to control point source VOC emissions. A new treatment method, using a packed bed of ceramic beads to thermally destroy contaminants, is also discussed. Figure 23 shows the general ranges of applicability for these options. The figure shows that most of these alternatives may be used over a range of concentrations that spans several orders of magnitude. Usually, however, each option is cost effective over a small part of that range. For example, GAC adsorption could be used to treat a vapor stream containing 10,000 ppm of hydrocarbon vapors, but the cost for carbon regeneration would be prohibitive. Rather, an incineration technique would likely be used. Likewise, an internal combustion engine could be effective at reducing a vapor stream containing

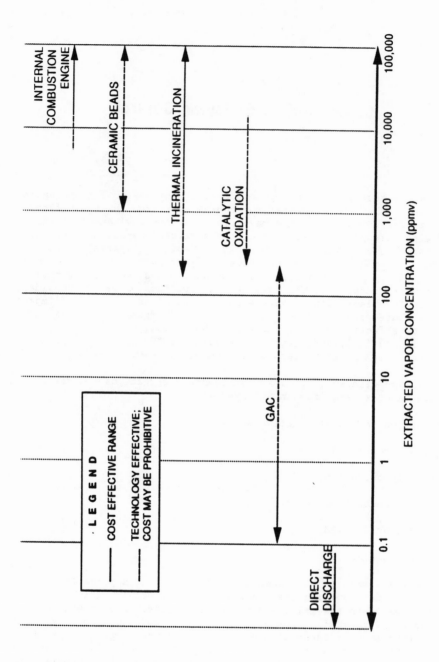

Figure 23. Applicability of Vapor Treatment Options

less than 1000 ppm, but this technology would require additional fuel and would probably not be the most appropriate vapor treatment choice.

GRANULAR ACTIVATED CARBON ADSORPTION

Adsorption refers to the process by which molecules collect on and adhere to the surface of an adsorbent solid (EPA, 1988b). This adsorption is due to chemical and/or physical forces. Physical adsorption (the more common type in this application) is due to Van der Waals' forces, which are common to all matter and result from the motion of electrons. Activated carbon is used as an adsorbent because of its large surface area, a critical factor in the adsorption process since the adsorption capacity is proportional to surface area. "Activated" carbon has significant surface area due to its internal pore structure; commercially available GAC typically has a surface area 1,000 to 1,400 m^2/gram.

Granular activated carbon (GAC) is the most common method of vapor phase treatment. GAC is popular for several reasons, including its relative ease of implementation and operation, its reputation as a commonly-used treatment, its ability to be regenerated for repeated use, and its applicability to a wide range of contaminants at a wide range of flow rates. Many vendors sell or lease prefabricated, skid-mounted units that can be procured and implemented in a matter of days. Carbon is economical only for relatively low mass removal rates, however; when the vapor concentration is high, carbon replacement or regeneration may be prohibitively expensive.

The adsorption capacity of the carbon depends on several factors, including influent vapor temperature and relative humidity and, most importantly, the influent VOC types and concentrations. Isotherms, which show the mass of contaminants that can be adsorbed per unit mass of carbon, are available to predict the contaminant-specific adsorption capacity for a specific type of carbon. Figure 24 shows isotherm data for benzene at various temperatures and pressures for a specific charcoal-based carbon. Isotherms for common compounds are available from carbon vendors. GAC generally has a high affinity for volatile molecules, such as hydrocarbons or chlorinated compounds, which are the most likely types of compounds to be removed via SVE; however, some hydrocarbons such as isopentane have relatively low adsorption capacities.

Although GAC has a very high surface area for adsorption of contaminants, the mass of contaminants removed via SVE may quickly exceed the carbon's capacity. At sites with high mass removal rates, due to high concentration, high flow rate, or both, the carbon may quickly become spent (i.e., the capacity is filled). Replacement and disposal of spent carbon can become very expensive, especially if the spent carbon must be treated as a RCRA waste.

An alternative to replacement of the carbon with off-site disposal or reactivation is on-site regeneration of the carbon. Such systems regenerate the carbon in place, using steam or hot air to desorb the contaminants. The contaminants recovered in liquid form may then be disposed or, in some cases, may be used as fuel to produce steam.

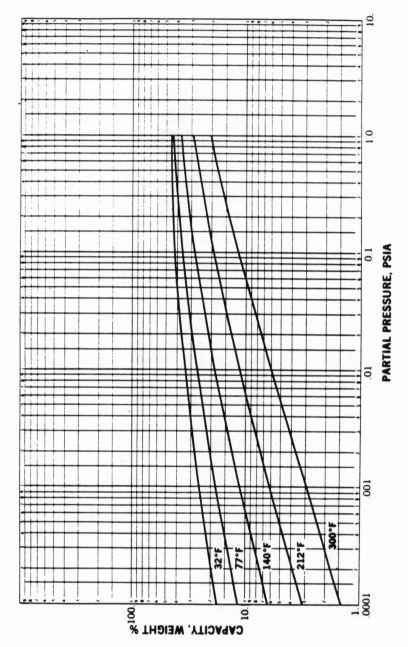

SOURCE: Calgon Carbon Corporation (Used with permission).

Figure 24. Benzene Isotherms

Connor (1989) states that this type of system has been able to achieve recovery of 36 gallons of product per day. That system has also shown that the steam generator unit requires about six gallons of fuel (gasoline) for every five gallons of product recovered. Thus, operating at maximum capacity, this unit would consume about fifty gallons of gasoline weekly for steam generation. This system becomes economical as the mass removal rate increases.

The relative humidity of the incoming vapor stream may severely limit the effectiveness of and increase the cost of the GAC. Water vapor will preferentially occupy adsorption sites, thereby decreasing the capacity of the carbon to remove contaminants from the air stream. For this reason, entrained water should be removed from the incoming vapor stream by use of an air/water separator. Vendors typically recommended that the vapor should be treated further to reduce the relative humidity to below 50 percent, usually by heating the air. Placement of the blower prior to the GAC unit is a way to utilize the heat of compression produced by the blower to increase the vapor temperature and thereby increase water's solubility in air, reducing the relative humidity. The GAC may be placed prior to the vacuum pump or blower to create a greater negative pressure through the GAC and to prevent overheating the GAC unit, although adsorption may be reduced under vacuum.

The heat generated by pumping systems and by the compression of vapors often results in an exhaust stream of elevated temperature. The off-gases from some vacuum systems must be cooled from 200 to 80 degrees F for efficient treatment before entering the carbon adsorption units. Figure 24 shows how the adsorption capacity decreases as temperature increases. Cooling may be accomplished by vertical, chilled water coils. Condensate generated during vapor cooling must then be treated.

Information on GAC design parameters is available from the carbon vendors. Calgon Carbon Corporation (Pittsburgh, PA) and Carbtrol Corporation (Westport, CT), among many others, supply pressure drop curves for the various GAC types they supply. The pressure drop curves are developed as a function of flow rate. Standard size containers are available depending on the expected air flow rate. Many vendors supply modular, prefabricated GAC units of 200 to 2000 pounds of activated carbon that may accommodate flow rates from below 400 scfm to over 1,000 scfm.

THERMAL INCINERATION

Several types of incineration options may be used to destroy vapor-phase contaminants resulting from SVE. Incineration ideally converts compounds to carbon dioxide and water. Complete destruction of contaminants requires very high temperatures, typically 1000 to 1400 degrees F. The destruction of the contaminants is a major advantage of this technique over carbon adsorption, which serves only to concentrate the contaminants onto the carbon, which must then be disposed.

Thermal incineration is the most basic of all incineration techniques. The vapors are heated to a very high temperature, usually in a combustion chamber, although some methods use a direct flame. To reach the required temperatures, the influent vapor is preheated and then enters the combustion unit.   In cases where the influent concentration is very high (on the order of percent by volume), the incineration may be self-sustaining. Fall and Pickens (1989) reported that vapor concentrations in the range of 50,000 to 90,000 ppm sustained a flame. Concentration of volatiles in the air stream might be increased by intermittent blower operation (the "pulse method") or by intermittently operating different extraction vents (Hutzler et al., 1989). For the case cited above, the flame was self-sustaining only when the system was operated eight hours per day; if operated continuously, a stable flame could not be maintained. For most vacuum extraction applications, the vapor concentrations are not high enough to sustain a flame and a supplemental fuel must be used to maintain the necessary temperatures. Natural gas or propane normally serves as the supplemental fuel.

For safety reasons, influent concentrations are normally limited to 25 percent of the lower explosive limit (LEL) (EPA, 1986). The LEL for gasoline is between 12,000 ppm and 15,000 ppm, depending on the gasoline's grade (A.D. Little, 1987) (Figure 25).

Direct incineration is not appropriate for influent vapor streams containing chlorinated compounds. Partial or incomplete combustion of chlorinated compounds could result in the production of chlorine gas and other PICs (products of incomplete combustion).

CATALYTIC OXIDATION

Catalytic oxidation is a variation of thermal incineration. In this process, the vapor stream is heated and passed through a combustion unit in contact with a catalyst. The catalyst unit is generally a metallic mesh, ceramic honeycomb, or packed bed consisting of catalyst-impregnated pellets (EPA, 1986). The catalyst is typically composed of a precious metal formulation (e.g., palladium or platinum) that facilitates the transformation of the contaminant molecules into carbon dioxide and water. Trowbridge and Malot (1990) describe a catalytic oxidation unit with a non-precious metal catalyst that may be used for chlorinated air streams. Figure 26 shows a schematic diagram of a catalytic incinerator unit.

In this process, a catalyst accelerates the chemical reaction without undergoing a chemical change itself. The catalyst increases the incineration reaction by adsorbing the contaminant molecules on the catalyst surface. The higher concentration of reactive materials serves to increase the reaction rate, thereby facilitating the oxidation process (Hardison and Dowd, 1977). Careful monitoring of extraction gas concentration is required, and the air stream must be diluted to be kept below 25 percent LEL (3,000 ppm) since higher concentrations may cause overheating, resulting in the catalyst melting.

The main advantage of catalytic incineration versus thermal incineration is the much lower temperature required with a catalyst. These systems typically operate at 600 to 900 degrees F (CSM Systems, 1989), versus temperatures of 1400 degrees F or higher for thermal incineration. The lower temperature results in lower fuel costs. Natural gas or propane are typical fuels used for vapor streams that do not contain sufficient heat value for a

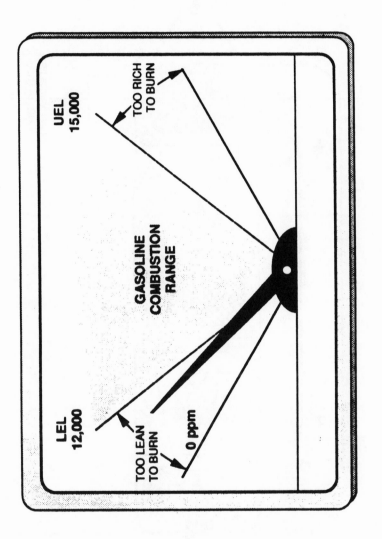

Figure 25. Explosimeter Readings

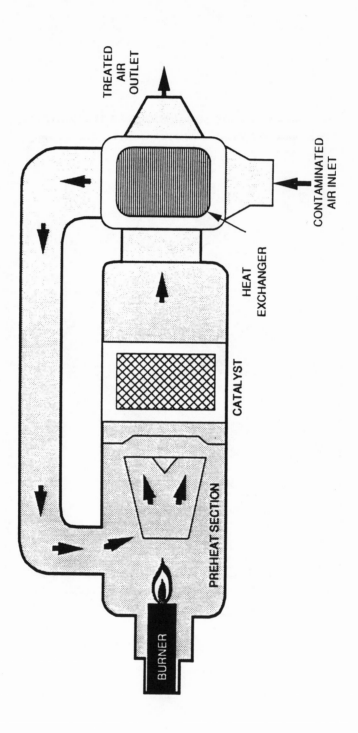

Figure 26. Catalytic Oxidation Schematic

self-sustaining incineration. Energy costs can be further reduced by
reclaiming heat from the exhaust gases, using the exhaust to preheat the
influent vapor stream, as shown in Figure 26.

Catalytic oxidation units require careful monitoring to prevent
overheating of the catalyst, which would result in its deactivation of the
catalyst. While the exact limitations depend on the heat value of the
influent vapor stream, concentrations over 3000 ppm VOCs are normally diluted
with ambient air to control the temperature in the catalytic unit. Safety is
also a concern with these units, as with any incineration method. Maximum
permissible Total Hydrocarbons (THC) concentrations vary by site, but are
usually below 25 percent of the LEL. THC concentrations are measured before
operation of the catalytic unit to determine the necessity of sidestream
dilution of the vapor stream. New technologies potentially capable of
treating chlorinated compounds by catalysis are currently under development
(Trowbridge and Malot, 1990).

INTERNAL COMBUSTION ENGINES

Internal combustion engines (ICEs) have been used for years to destroy
landfill gas. The application of this method to hydrocarbon destruction is
recent, with the first operational unit having been installed in 1986.
Currently, over one hundred of these units are operating in southern
California and providing good destruction and removal efficiencies.

The internal combustion engine used for this technique is simply an
industrial or automotive engine with its carburetor modified to accept vapors
rather than liquid fuel. Virtually any make of engine can be used:
Volkswagen, Audi, Ford, Chevrolet and others have all been reported as having
been used. The size of the engine (expressed in cubic inches) reportedly
greatly affects the flow rate of air through the engine, with larger capacity
engines able to handle larger flow volumes. For example, RSI (Oxnard, CA)
sells an internal combustion device that uses a 1988 Ford 4-cylinder, 140
cubic inch engine, which is rated to accept 30 to 60 scfm of inlet flow. By
contrast, Wayne Perry Inc. (Buena Park, CA) employs a 460-inch engine that
accepts 400 scfm. RENMAR, a manufacturer of self-contained ICE system,
reports a correlation of roughly 100 scfm flow-through per 300 cubic inches of
engine capacity without loading the engine (i.e, running it at idling speed).
This flow rate could be increased greatly by loading the engine (i.e., forcing
the engine to work harder and do more work by, for example, generating
electricity with the unused work potential of the engine). Differences in
destruction efficiencies based on engine type are unknown at present.

A second required modification to the engines is the addition of a
supplemental fuel input valve. When the intake hydrocarbon concentration is
too low to sustain complete combustion, a supplemental fuel source must be
added to ensure complete combustion. Propane is the fuel used almost
universally, although one vendor reported that tests with natural gas showed
greatly reduced (by 50 to 75 percent) energy costs. The concentration below
which supplemental fuel needs to be added is uncertain at this time. The
engines are also equipped with a valve to bleed in ambient air to maintain the
required oxygen concentration. Because soil vapor may have very low

concentrations of oxygen, especially during the initial stages of operation, ambient air is added to the engine, via an intake valve, at a ratio sufficient to bring the oxygen content up to the stoichiometric requirement for combustion. One vendor suggested a ratio of 4 parts ambient air to one part extracted vapor at start up, with the required ratio of ambient air decreasing as the soil oxygen content increases. This requirement for bleed-in air reduces the effective flow rate that the engine can treat; for example, at a flow rate of 100 cfm and a bleed-in ratio of 3:1, only 25 cfm would be extracted soil vapors, while 75 cfm would be ambient air.

A catalytic converter is an integral component of the system, providing an important polishing step to reach the low discharge levels required by many regulatory agencies. A standard automobile catalytic converter, using a platinum-based catalyst, is normally used. Data from the South Coast Air Quality Management District (SCAQMD), the air quality regulatory body for Los Angeles and the surrounding area, show that the catalyst reduced concentrations of total petroleum hydrocarbons (TPH) from 478 ppm to 89 ppm and from 1250 ppm to 39 ppm, resulting in important additional contaminant removal (Millican, 1989). SCAQMD requires a catalytic converter to permit this type of system. Catalysts have a finite life span (typically expressed in hours of operation) and must be monitored as that time approaches to ensure that the catalyst is working properly. The length of operation of the catalyst depends on the vapor concentration, whether lead is present, and the amount of propane added. A range suggested by one equipment vendor was 750 hours to 1500 hours (about one to two months) of operation. Upon deactivation, they can be replaced easily with any automobile catalytic converter, available at most auto parts stores.

To date, the use of ICEs appears to be limited to California, mostly in the South Coast Air Quality Management District in southern California, which has among the most stringent air discharge regulations in the country. SCAQMD has permitted over 100 ICEs for use in their district. RSI, Inc. (Oxnard, CA) has installed more than thirty ICE systems, all in California.

Data obtained from ICE operators and regulators, summarized in Table 7, show that ICEs are capable of destruction efficiencies of well over 99 percent. They have been especially useful in radically reducing incoming vapor streams with very high concentrations of TPH (up to 30 percent volume) to levels below 50 ppm. Results of tests for specific compounds (BTEX) show not detected in some cases and below 1 ppm in many other situations. The total destruction capacity may be expressed as mass removal rate. One ICE operator reported a mass removal and destruction rate of over one ton per day (about 12 gallons per hour) (Perry, 1989).

No information has been obtained to indicate how the destruction efficiency varies with changes in influent type and concentration, engine type and operation, the effects of manual versus automatic control, and the effects of physical parameters like relative humidity, temperature, contaminant concentration, and other variables. Landfill gas experience may shed light on some of these issues.

TABLE 7.   DESTRUCTION EFFICIENCIES OF ICEs

Parameter	Initial Concentration (ppm)	After Catalytic Converter (ppm)	Removal Efficiency (%)	Reference
THC	38,000	89	99.76	Millican, 1989
	200,000	39	99.98	
THC	318,832	16 ppm	99.99	Wayne Perry, 1989
Benzene	995	ND (<10 ppb)	99.99	" "
Ethylbenzene	125	ND (<10 ppb)	99.99	" "
Toluene	1005	0.014	99.99	" "
Xylenes	1550	<11.5 ppb	99.99	" "
TPH (non-methane)	49,265	225	99.56	RSI, 1989
Methane HCs	741	109	85.29	" "
Benzene	380	0.8	99.79	" "
Toluene	400	1.1	99.73	" "
Xylenes	114	0.7	99.39	" "
Ethylbenzene	18	<0.5	--	" "
TPH	65,450	30	99.95	Rippberger, 1989
	34,042	14.5	99.96	" "
	30,500	1.4	99.99	" "
	39,000	4.7	99.99	" "
Benzene	1,094	67	93.88	Rippberger, 1989
	470	1.6	99.66	" "
	785	0.63	99.92	" "
	730	0.056	99.99	
THC	58,000	160	99.72	RSI, 1989
Benzene	1,400	0.13	99.99	" "
Toluene	720	0.024	99.99	" "
Ethylbenzene	77	0.062	99.92	" "
Total Xylenes	320	0.13	99.96	" "
THC	26,000	140	99.46	" "
Benzene	960	0.024	99.99	" "
Toluene	840	0.020	99.99	" "
Ethylbenzene	91	ND (0.02)	100.00	" "
Total Xylenes	360	0.080	99.98	" "

The use of internal combustion engines as a vapor treatment device for extracted soil vapors may possess several advantages over conventional treatment methods (carbon, thermal oxidation, or catalytic oxidation), at least for certain instances. Perhaps the most important advantage that this method has over other vapor treatment methods is that the ICE produces an easily harnessed source of power that can be used to produce useful work. In fact, some vendors sell self-contained units that use the ICE to power the vacuum pump that is the driving force behind vapor removal. An added benefit of this system is that vapors cannot be extracted unless treatment is also occurring, eliminating the possibility of vapors bypassing the treatment system. RENMAR, an equipment supplier, contends that the vacuum extraction system consumes only about 25 percent of the useful work produced by the engine. Other ideas for using the power could include lighting the site, heating a field trailer, or similar ideas. The engine could also be used as injection air for the SVE system.

The ICE is able to handle very high concentrations of extracted air, such as would likely be found over a large free product plume. Reports of inlet vapor concentration have gone as high as 300,000 ppm (Millican, 1989). By contrast, catalytic incinerators and thermal incinerators sometimes require dilution with bleed-in air to a safe level, reducing their upper limit. ICEs can also accept fairly low concentrations (to below 1000 ppm), although supplemental fuel use increases and the cost effectiveness decreases greatly at reduced intake concentrations. Further, the removal efficiency compares favorably with other treatment methods, based on the limited data available from actual system installations.

Another advantage of ICEs is their portability. Typically, the self-contained units are skid-mounted or put on a trailer and can go from site to site very easily. The site requirements may also favor ICEs over other oxidation methods. ICE units are reportedly smaller and less noticeable than direct thermal incineration units and may be more appropriate for areas that wish to remain low profile.

Some disadvantages of ICEs have been noted. The primary drawback may be that the method requires a fairly high degree of manual supervision, especially when the system is being started up. Mainly, the air to fuel ratio must be adjusted to maintain the proper conditions for complete combustion. Microcomputers are available to monitor and adjust the air to fuel ratio and add propane as needed; however, immediately after system startup, the characteristics of the extracted vapors may change so quickly that manual adjustment is required. As the system operates for a longer period, manual attendance may no longer be required.

Another potential disadvantage of this technique is the somewhat limited flow rates that can be removed from the soil and treated. The flow rate is limited especially by the ambient air that must be added to maintain required oxygen levels. The use of oxygen, rather than air, might increase the relative proportion of extracted vapors that the engine treats. A third potential disadvantage of ICE is the noise level. Users have reported that these devices are quite loud and their use may be restricted in certain neighborhoods.

PACKED BED THERMAL PROCESSOR

A technology that may have application to the control of vapor emissions
is the packed bed thermal processor. This system consists of a bed packed
with ceramic beads that are heated to 1800 degrees F, through which the vapor
stream passes and the contaminants are destroyed. The patented packing
geometry and the uniformly high temperature that the ceramic beads are able to
result in nearly complete destruction of the influent vapors, without the use
of a flame.

IN-Process Technology (Sunnyvale, CA) has used these systems to destroy
vapors at several chemical and other industrial plants. The company is
currently investigating its applicability to the remediation market
(Fredricks, 1990). According to Fredricks, this technology has several
characteristics that may allow it to be used successfully as an emission
control system for remediation streams. First, the removal efficiency is
extremely high and very reliable. Tests have shown efficiencies of 99.99+
percent, and this removal is attained continuously. The residence time can be
adjusted to attain any necessary removal efficiency. Other incineration
techniques use a flame, which leads to non-uniform heating and non-uniform
removal efficiencies. This method does not use a flame, and the geometry of
the packed ceramic beads results in complete, uniform combustion, so that no
NO_x compounds are produced. Once the beads reach the proper temperature of
1800 degrees F, which is reached by electrically heating the ceramic, no
additional energy input is required if the heat value of the vapors is
sufficient. This point is near a concentration of 2000 ppmv. If the
concentrations are below this value, natural gas or propane can be bled in
with influent to maintain the proper temperatures.

As with any incineration technique, excess air needs to be added to
dilute the concentration to safe levels if the influent is too rich. This
method has handled concentrations at the percent level. Throughput levels
depend on the model selected (presently 100, 200, and 500 scfm) with higher
flow rates met by combining two or more of the modular units.

Perhaps the greatest advantage of this technique, according to the
vendor, is its ability to destroy chlorinated compounds without the production
of hazardous by-products and without degradation of the ceramic beads. This
ability sets this method apart from most other incineration techniques.

BIOTREATMENT

Biofilters have been used for many years to treat odors (Carlson and
Leiser, 1966; Prokop and Bohn, 1985; and others). Typically, these soil beds
are used to control malodorous gases resulting from sewage treatment plants or
industrial plants. Biofilters have also been used recently to treat vapor
phase VOCs prior to atmospheric discharge. Pilot studies indicate that
significant VOC removal may be possible with this method (Johnson, M., 1989).
This process operates by introducing the VOC-laden vapor to a soil bed, which
serves as a growth medium for microorganisms. As the contaminants flow up
through the soil they sorb onto the soil surface, where they are degraded by
microbes. The process whereby the contaminant adsorbs onto the soil is the

same process that occurs during GAC adsorption. Soil has a much smaller surface area than activated carbon and therefore has a smaller adsorption capacity; however, the degradation activity of the soil microbes serves to oxidize the contaminants and allow further contaminant sorption at those sites. The theory behind biofilters is analogous to biological wastewater treatment. That is, by providing an environment with suitable amounts of oxygen, nutrients, pH, and temperature, microbiological degradation will occur.

Soil biofilter construction is relatively straightforward. A network of perforated piping is buried in the soil bed. The exhaust gas is pumped through the piping network, from which the gas flows up through the soil. Adsorption and subsequent degradation occur on the soil surface. The soil bed must be sufficiently porous to allow large volumes of the exhaust air to move expeditiously through the bed; suitable media include sandy loam soils or mature compost (Johnson, M., 1989). The beds must be designed to avoid short circuiting due to drying and cracking. Clay or organic matter serve to give the soil matter sufficient sorption sites. Aerobic conditions must be maintained to permit oxidation of the organics and survival of the microbes. Aerobic conditions are maintained by ensuring that the exhaust air contains sufficient oxygen. The air permeability must be maintained at a level that permits the discharge of the vapor extraction flow rate. The moisture content plays a crucial role in biofilter operation. Soil moisture should be maintained at a level to maximize biological activity. A leachate collection system may be necessary where large quantities of water are being added. Neff (1989) listed several other operational parameters and their optimal ranges: retention time (>15 seconds); temperature (15 to 45 degrees C); pH (7 to 8); moisture content (50 to 70 percent by weight); media porosity (80 to 90 percent); and influent gas relative humidity (60 to 80 percent).

The advantages of the soil bed treatment system are the low cost and the complete destruction of the contaminants. A disadvantage is the acclimation period required and the potential for system upset following changes in influent concentration. This type of system is widespread in deodorization applications, and it is now being applied to VOC control. Experience will tell whether this technology will be applicable to remediation sites.

DIRECT DISCHARGE

In some cases treatment of the extraction well vapor stream may not be required and direct discharge of vapors to the atmosphere may occur. The concentration of the contaminants, the flow rate of the SVE system, and the presence or absence of nearby receptors are generally taken into consideration when evaluating direct discharge options.

7. Cost of Soil Vapor Extraction

INTRODUCTION

The costs for a soil vapor extraction system may be apportioned into three general categories. Site investigation costs include site history review, site assessment, and pilot testing. Capital costs consist of costs for procuring and installing the system components, as well as design and engineering fees, permitting costs, and contingencies. Operation and monitoring (O&M) costs are those associated with the continued operation of the system, including power and labor, system monitoring, and clean up attainment analytical costs.

SITE INVESTIGATION COSTS

The basic costs to perform a site investigation include sampling and monitoring equipment, laboratory analysis, personal protective equipment (PPE), and labor. These costs are discussed below for each segment of a site investigation.

Site History Review

During the site history review, all available data pertinent to the site should be located and collected. This data normally includes maps from the U.S. Geological Survey and the Soil Conservation Service, aerial photographs, site plans, and operational records. Interviews with persons familiar with the site are often helpful.

The cost of the site history review will depend primarily on the labor expended to obtain the information and, to a lesser extent, the cost associated with obtaining maps or other information. The time spent on the initial site investigation may range from only one or two days for a service station UST release that has a clearly identified leak history (e.g., a vapor monitor detects a leak that inventory records indicate began two weeks earlier) to one month or more, continuing even as field work begins, for a larger, more complex site (a tank farm or a manufacturing facility). The cost of map procurement may range from nil to over $1,000 if original aerial photography is required. The total cost of the initial site investigation may range from $5,000 to $20,000.

Preliminary Site Screening

The objective of a preliminary site screening is to assess, rapidly and cost-effectively, the nature and extent of contamination. During a site survey, key features of the area (such as the presence of sewers and utility

97

lines) are noted, and health and safety conditions for the site are determined.   A soil gas survey and groundwater sampling are often performed to gain additional data.   Sampling of soil, soil gas, groundwater and air normally occurs in this step.

The preliminary site screening requires, at a minimum, an organic vapor monitor to identify total VOCs and a combustible gas indicator (CGI) to identify combustible gas concentrations and oxygen-deficient atmospheres.   The Foxboro OVA (organic vapor analyzer) 128 GC and the HNu PI 101 model are the most commonly used models.   The OVA model uses a flame ionization detector (FID) while the HNu uses a photo ionization detector (PID).   It is usually beneficial to incorporate both an FID and a PID vapor monitor in the survey, because these instruments respond differently to various gases and environmental conditions (e.g., the FID is sensitive to methane, while the PID is not).

This preliminary site screening is usually conducted in level D safety equipment, which requires a minimum of personal protection equipment (PPE), unless otherwise indicated.   However, the first step to occur during this screening is to monitor for vapors that may be immediately dangerous to life or health (IDLH), other conditions that may result in death, such as explosive vapor concentrations, and oxygen deficient atmosphere. Containers, drums, spills and other potential safety hazards should be identified at the beginning of the site screening.

Subsequent tasks might include subsurface investigations using ground penetrating radar (GPR), metal detection, and soil gas surveys.   At this stage of the site investigation, the soil gas survey should be rapid. Laboratory GC equipment should be avoided until the detailed site characterization.   Portable field monitoring equipment such as the OVA and the Photovac 10S50 are suitable for the soil gas survey if both quantitative and qualitative data are required.   Soil gas systems are also available from other vendors such as K.V. Associates, Inc. and Xitech.

The total cost for the preliminary site screening will depend on the cost for labor and mobilization, monitoring equipment, personal protective equipment, sampling equipment and supplies.   Laboratory analytical costs should be negligible during this stage of the site investigation because analyses should be done primarily in the field.   The labor cost will depend on the number of personnel required, site conditions (both size and ease of access), weather, and the level of personal protection required (significantly more time is required for greater levels of protection).

The cost for monitoring equipment will depend on the type of equipment used and whether it is purchased or rented.   At a minimum, a vapor monitor and CGI are required.   The cost for this equipment is presented in Table 8.   The costs range from $4600 for the HNu 101 with a calibration kit to $7,300 for the Foxboro OVA 128 GC.   This equipment can also be rented at a cost of approximately $350 to $550 for a ten day period or $900 to $1,200 per month (CAE Instrument Rental, Inc.).   The CGI may be purchased for $650 to $1,950, or can be rented for $50 to $220 for a ten day period.

TABLE 8.   MONITORING EQUIPMENT COST ESTIMATES

Model	Vendor	Range (ppm)	Cost ($)	Notes
OVA-128	Foxboro	0-1000	6100	FID
OVA-128GC	Foxboro	0-1000	7300 1125/month	w/GC rental
OVA-108	Foxboro	1-10000	6100	FID
OVA-108GC	Foxboro	1-10000	7300 1125/month	w/GC rental
HNu PI-101	HNu Systems	1-2000	4250 900/month	rental
Micro TIPII	Photovac	0.1-2000	4400 900/month	PID w/data logging rental
OVM 580A	Thermo Env.	0.1-2000	4700 935/month	PID rental w/ data logging
HNu-311	HNu Systems		14000	GC/PID
10S50	Photovac		15500	GC/PID
10S70	Photovac		18500	GC w/data logging
Sniffer 302	Bacharach		940	O_2/LEL
1314	GasTech		1100	"
Sentinel 4	Bacharach		1450	Personal Monitor

NOTE: Costs are estimates (1989). Contact vendor for actual price quotes.

The cost for PPE varies depending on the level of protection (Table 9).
For example, a suit for Level B costs $100, while the Level A suits cost from
$440 to $4,000, depending on the type of suit required. The NIOSH Pocket
Guide to Chemical Hazards (U.S. Department of Health and Human Services, 1985)
is available to determine the appropriate PPE based on the contaminants
present. Organic vapor monitor badges, which may be worn during the on-site
work, can be purchased for $125 (box of 10).

TABLE 9. WORKER PROTECTIVE EQUIPMENT COSTS

Type	Vendor	Level of Protection	Costs ($)	Notes
Suit				
tyvek	Dupont	C		not water resistant
saranex	Dupont	B		water resistant
Top security suit	Kappler	B	100	
Frontline Suit	Kappler	B	115	chemical resistant
Responder Encapsulating	Lifeguard	A	440	chemical resistant
Responder Encapsulating	Lifeguard	B	160	chemical resistant
Aluminum PBI/ Kevlar			660	fire resistance/outer cover
Butyl	Conners Env.	A	1,450	general purpose-chemical
Viton	Conners Env.	A	3,300	PCBs, chlorine, hydrocarbons
PVC Chemical	Conners Env.	A	1,350	chemical hazards
Teflon/Nomex Encapsulating	Lifeguard	A	4,000	chemical resistant
Gloves			(pr.)	
Viton	Conners Env.		37-47	PCBs
Butyl	Conners Env.		18-25	gas or water
Silver Shield	Conners Env.		3	disposable
Neoprene	Pioneer		6-16	PCBs 22 to 30 min
Argus	Playtex		7.50	PCBs, oils, acids
PVA	Edmont		24	aromatic and chlorinated solvents
Latex	Conners Env		13/50 pair	disposable
Rubber	Edmont Environmental		18/dozen	acids, alkalis, ketones
Boots				
PVC	Conney		13-23	Waterproof, Chemical resistant
Neoprene	Conney		53	Chemical & petroleum resistant
Hazmex Latex			7	"chicken boots"
PVC Shoe/Boot Protector			3.50	
Respirators				
Half-Mask Facepiece	North	C	21	
Full Facepiece	North	C	175	
Cartridges	North	C	25/box of 6	organic vapor
Powered air Purifying Respirator	North	C	560	requires cartridges
Pressure Demand Airline Respirator	North	B/A	620	
Continuous Flow	North	B/A	400	
SCBA	North	B/A B/A	1,725 2,350	30 min 60 min
Air Pump	Allegro Allegro	B/A B/A	850 2,400	2 min 6 min

Detailed Site Characterization

Additional site characterization geared specifically toward SVE is normally undertaken if the preliminary screening results indicate that SVE may be an appropriate remedial technology. This step includes more detailed soil and contaminant analyses to determine specific parameters to be used during remedial design.

Soil samples may be collected by a hand auger, auger drill, split spoon sampler, or Shelby tubes. These samples can be used to determine various soil properties such as the permeability, porosity, bulk density, moisture content, pH, cation exchange capacity (CEC), and organic carbon content. In situ measurements of hydraulic conductivity and infiltration rate can be made with a Guelph Permeameter and an infiltrometer, respectively. The determination of other soil parameters, such as residual saturation and leaching characteristics, are possible with soil column testing apparatus. Groundwater samples are taken via monitoring wells or mini-well points.

Table 10 lists costs for soil, groundwater, soil gas, air and other environmental measurement equipment. Costs on this table are estimates and the vendors should be contacted for actual equipment costs. Table 11 lists typical analytical costs for various analyses. Geological evaluations are required to characterize media at depths below about 15 feet, the rough limit of hand equipment, or into shallow rock structures. Common geologic sampling equipment includes split-spoon samplers and Shelby tubes. Soil and rock samples should be evaluated for texture, density, organic carbon content, and contaminants. Samples may be "screened" by headspace analysis on field GC equipment in lieu of laboratory analysis.

Geologic cross sections, which will be used in the evaluation of contaminant migration pathways, can be developed at the site from boring logs. Geophysical techniques may be required to evaluate specific subsurface conditions. These techniques may include seismic refraction and reflection, magnetometry, electromagnetics (EM), resistivity and gravity. Seismic measurements can be performed to determine the thickness and number of subsurface layers, depth to bedrock, and the presence of fractures and cavities. Magnetometry detects the presence of buried ferrous metals and is normally used to detect drums. Electromagnetic measurements can be made to assess lateral variations in soil and rock, such as fractures and karst features, as well as to locate shallow drums. Resistivity measurements are conducted to evaluate the depth and thickness of soil and rock layers and the depth to the water table. Gravity determinations are used to map major geologic features over a large area or for detecting local fractures and cavities.

The cost for the geologic investigation depends on the costs for drilling and geophysical analyses. Well costs are approximately $300 to $1,000 for deployment, $20/foot of drilling, $75 for decontamination equipment (steam cleaner), $100/hour for decontamination (labor), and $50 per sample (split-spoon). Geophysical equipment can be purchased from vendors at prices shown in Table 10. The conductivity meter for shallow depths (less than 20 feet) costs $12,200, while the meter for greater depths costs $18,200. A

TABLE 10. SITE INVESTIGATION EQUIPMENT COSTS

SOIL

Equipment	Vendor	Cost($)	Notes
Auger (manual)	AMS (Forestry Suppliers)	100	
Auger (power)	Little Beaver	1225	11 hp
	Hoffer PH980	360	85 cc
Soil core sampler	AMS	250	w/slide hammer
Retaining cylinders	AMS	10	stainless steel
Split-spoon sampler	SoilTest Inc.	500	stainless steel
Thin-walled tubes	SoilTest Inc.	800/doz	stainless steel

GROUNDWATER

Equipment	Vendor	Cost($)	Notes
Bailer (economy)	Forestry Suppliers	55	350 cc
Bailer (PVC)	Forestry Suppliers	60	36"x1.66" O.D.
Bailer (stainless steel)	Forestry Suppliers	90 - 140	Johnson
Bailer (teflon)	Forestry Suppliers	130 - 220	
Pump (portable)	Masterflex	500 - 560	peristaltic pump
Tubing	Masterflex	44/roll	silicon, roll of 25'
Pump (portable)	Geoguard	4000	bladder pump
Dedicated well sampler	Geoguard	600 - 700	

SOIL GAS

Equipment	Vendor	Cost($)	Notes
Gas sampler system	Xitech/Vista 4000	2500	also groundwater
Gas sampler system	K-V Associates/Macho	2575	
Gas sampler system	K-V Associates/Basic	1675	
Gas sampler system	K-V Associates/Hefty	3600	

AIR

Equipment	Vendor	Cost($)	Notes
Sampling pump	SoilTest	330	electric
Sampling pump	SoilTest	35	manual

ENVIRONMENTAL MEASUREMENTS

Equipment	Vendor	Cost($)	Notes
Double ring infiltrometer	SoilTest	1550	infiltration rate
Permeameter	Soilmoisture Guelph Permeameter	1100	soil permeability
Conductivity meter	SoilTest	12200 - 18200	197' soil; <20' rock
Resistivity meter	SoilTest	2715	<100 meters
Resistivity meter	SoilTest/I.P. System	12800	induced polarization
Seismograph	SoilTest	4250	signal enhanced
Groundwater Quality	YSI 33 S-C-T	500	
Groundwater Quality	YSI 3560	2000	
ORP electrode	Orion	100	
CI electrode	Orion	375	

TABLE 11.   TYPICAL ANALYTICAL COSTS

Analysis	Soil/Sediments	Groundwater	Soil Gas
VOCs	$225	$200	$250
ABNs	500-550	525	
TPH	75-125	100	
BTX	180	150	
Pesticide/PCBs	175	150	150
Texture	75- 125	--	--
Moisture	10 - 20	--	--
pH	5 - 15	10	--
CEC	75 - 85	--	--
Organic Matter	10 - 40	--	--
Permeability	335 - 495	--	--

resistivity meter and an automatic resistivity-induced polarization system are sold for $2,715 and $12,800, respectively, by Soil Test. A signal-enhanced seismograph costs $4,250.

Characterization of Contaminants

    A contaminant characterization helps to determine the applicability of SVE by identifying the compounds present and their concentrations. A Quality Assurance/Quality Control(QA/QC) sampling plan, addressing soil, groundwater, and soil gas sampling should be in place to ensure data precision and representativeness. Wastes (e.g., lagoons), surface water and sediments should also be sampled if appropriate.

    The soil should be analyzed for VOCs, acid/base/neutrals (ABNs) or semi-volatile organic compounds, and total petroleum hydrocarbons (TPH). VOC analysis will determine the concentration of volatile compounds that may be amenable to SVE. EPA Method 8240 is used to evaluate VOCs. Targeted analyses for specific, indicator compounds such as benzene, toluene and xylenes will reduce analytical costs. Analysis of the soils for ABNs (EPA Method 8270) will identify less volatile compounds that are less likely to be removed by SVE. ABN analysis may be augmented by the less expensive TPH analysis (EPA method 418.1), which does not identify specific compounds, but gives only a total of the petroleum compounds present in the soil. Table 10 lists sampling equipment costs and Table 11 lists typical analytic costs. These constitute most of the costs of this part of the site investigation.

    During this phase, groundwater analyses may be required. Bladder pumps are preferred over bailers for collecting groundwater samples, although pumps are more expensive. Geoguard, for example, sells both dedicated and portable bladder pump systems. The portable pump system (pump, compressor, controller, and 100 feet of tubing) costs $4,000. The portable compressed air source for dedicated sampler systems costs $300. The cost per well for the sample is $600 to $700. Glass sampling bottles will cost approximately $3 each and vials for VOA are $2.50 each.   Laboratory analysis for groundwater contaminants will cost $200-250 for VOCs (EPA Method 624), $400-500 for ABNs (EPA Method 625), $75-100 for TPH (EPA Method 418.1), and $150 for pesticides/PCBs (EPA Method 608).

CAPITAL COSTS

        Capital costs for SVE systems include the cost of procuring and
installing all the equipment, piping, and instrumentation. Design and
engineering fees, permitting costs, and contingencies are also included under
capital costs. Many different components may comprise a complete system
(Figure 18), but in general the capital costs can be divided into three main
groups:

   o   Vapor capture, including costs associated with extraction and
       injection well installation, impermeable surface seals, and
       groundwater level control;

   o   Vapor removal, including costs associated with getting the vapors
       out of the ground and to the vapor treatment device and including
       the costs of the vacuum pumps/blowers, piping, valves, mufflers and
       monitoring equipment; and

   o   Vapor treatment, including vapor pretreatment (air-water separator),
       sidestream treatment, air conditioning, and the vapor treatment
       device.

Within each category, some components may not be necessary for every
site; for example, groundwater depression pumps would be used only at sites
where the contamination exists at or near the water table. Vapor treatment is
normally used only where the discharge limits are above allowable levels.

        The costs presented below were developed from a vendor survey conducted
in October, 1989 and are typical of those typically encountered by SVE users.
These costs generally do not include delivery, installation, and trouble-
shooting (unless indicated), which may add significantly to the overall cost.
Appendix C is a partial listing of vendors who supply SVE equipment. No
judgment of any vendor should be inferred or implied based on inclusion or
exclusion from this list.

Extraction Wells

        Extraction wells are normally constructed from schedule 40 PVC
(polyvinylchloride) piping of various diameters (2" to 12"). Polypropylene
(PP) or chlorinated polyvinyl chloride (CPVC) are more rigid and may be used
where stronger piping is required. A typical 30 feet deep extraction well
installation will usually cost from $2,000 to $4,000. Of this cost, materials
such as casing (riser), well screen, plugs, filter pack materials, bentonite,
and cement grout may range from $500 to $2,000 per well, depending on the
method of construction. Table 12 shows the range of costs for various
extraction well components. PVC piping, for example, costs as little as $2
per linear foot (lf) for 2" diameter casing to up to $12/lf for 6" diameter
casing. Similarly, PVC screens cost from $2 to $15 per linear foot depending
on diameter. Ball valves (PVC) cost $60 for a 2" riser to $300 for a 6"
riser.

Injection Wells

Injection wells and inlet wells are constructed similarly to extraction wells and have very similar costs. They are normally used to enhance vapor flow or control vapor paths. Existing monitoring wells are often used as injection or inlet wells.

Impermeable Surface Seal

Impermeable surface seals may be used to control vapor flow paths toward the extraction wells and reduce the entry of atmospheric air in the area of the well. At sites where a surface seal is present, the radius of influence of the extraction wells will increase as vapors are drawn from greater horizontal distances, permitting the removal of contaminants from a larger volume of soil. Surface seals can be constructed of high-density polyethylene (HDPE) or low-density polyethylene, bentonite clay, or asphalt. Material costs depend most directly on the area covered and are normally expressed in dollars per square yard ($/yd^2). HDPE, which is often used in landfill closure, costs about $5.00/yd^2 for 40 mil thickness. Polyethylene (10 mil thickness) costs about $2.25/yd^2. Bentonite clay can also be used as an impermeable barrier. Costs range from $2.22/yd^2 for a 4" layer to $3.33/yd^2 for a 6" application. Asphalt paving (2" layer) costs about $9.24/yd^2. This material would likely be used only for sites that will remain paved, such as a service station. Often, industrial and commercial SVE operations are implemented at sites that are already paved. For all methods, installation costs may add considerably to the total overall cost.

Groundwater Level Control

Extraction wells are normally screened through the contaminated layer. At sites where the contamination exists at or near the water table (as is common for petroleum and other products that are lighter than water), the application of a vacuum in a well will cause upwelling of the water table, especially in silty or clayey soils. This may result in the extraction wells drawing up large quantities of water or free product, which must be removed from the vapor stream. To minimize liquid entrainment, pumping wells may be installed near the vapor extraction wells to lower the water table.

Where free product exists on the water table, product recovery equipment is usually employed. These devices often use water table depression as a means of capturing free product and thus, can be used in conjunction with extraction wells as a means of groundwater level control.

Groundwater depression systems are available from vendors such as R.E. Wright Associates, Inc., who sell the Auto Skimmer, and Del Harlow Enterprises, Inc., who sell the Reclaimer System. These systems are capable of both water level depression and recovery of floating hydrocarbons. The water level depression system marketed by R.E. Wright Associates, Inc. is capable of removing 45 to 95 gallons per minute (gpm) and sells for $3,700, uninstalled.

TABLE 12. SVE SYSTEM COMPONENTS
CAPITAL COSTS

COMPONENT	TYPE	SIZE	CAPITAL COSTS ($)	NOTES
EXTRACTION WELL CONSTRUCTION			20-40/FT	
CASING	PVC	2 IN	2-3/FT	SCH. 40 PVC
		4 IN	3-5/FT	
		6 IN	7-12/FT	
SCREEN	PVC	2 IN	2-4/FT	
		4 IN	5-7/FT	SCH. 40 PVC,
		6 IN	10-15/FT	ANY SLOT SIZE
SAND PACK			15-20/CU FT	
GRAVEL PACK			20/CU YD	
PIPING	PP	2 IN	1/FT	
		4 IN	3.50/FT	
		6 IN	8/FT	
	PVC	2 IN	1/FT	SCH. 40 PVC
		4 IN	3/FT	
		6 IN	5.25/FT	
	CPVC	2 IN	4/FT	SCH. 80 PVC
		4 IN	12.50/FT	
		6 IN	24/FT	
				vendor- M&T Plastics
VALVES (BALL)	PVC	2 IN	65	SCH. 40 PVC
	single union	4 IN	300	2in & 4in threaded socket
		6 IN	700	6in flange end connection
				M&T Plastics
JOINTS (ELBOW)	PVC	2 IN	2.50	SCH. 40 PVC
	90 degrees --slip	4 IN	16	threaded, socket
		6 IN	51	end connections
WATER TABLE DEPRESSION PUMPS			3700	R.E.Wright Assoc.
			3000	w/Explosion Proof Pump Motor Control System
SURFACE SEALS	BENTONITE 6 in		3.33/SQ.YD.	
	BENTONITE 4 in		2.22/SQ YD	
	POLYETHYLENE 10 mil		2.25/SQ YD	
	HDPE 40 mil		5/SQ YD	
	ASPHALT 2 in		9.24/SQ YD	
BLOWER (FAN)		1 hp	1700	Environ Instruments
		1.5 hp	2000	
		2 hp	2200	
		3 hp	2700	
		5 hp	3300	
		10 hp	5000	
		30 hp	6000	
	centrifugals blowers	2.5 hp	600	
		25 hp	12000	includes installation
		50 hp	42000	

(continued)

TABLE 12. SVE SYSTEM COMPONENTS
CAPITAL COSTS (continued)

COMPONENT	TYPE	SIZE	CAPITAL COSTS ($)	NOTES
AIR/WATER SEPARATOR			1500-2400	
	knockout pots	800 gal	11600	vendor-Water Resources
		20 gal	1470	installation 33%
		35 gal	1560	of capital costs
		65 gal	1750	
		105 gal	2150	
		130 gal	2350	
INSTRUMENTATION				
VACUUM GUAGE (MAGNEHELIC)			50-75	
FLOW (ANNUBAR)			300	
SAMPLING PORT	brass T		20-30	
CONCRETE PAD			450/CY	
HEAT EXCANGER	& housing unit		1400	
FIBERGLASS SHED	8'x10'		8500	
FLAME ARRESTOR	w/o ss element		665	vendor- Stafford Tech.
	w/ ss element		735-930	
AIR RELIEF VALVE			225	vendor- Stafford Tech.
SOIL GAS PROBE			30-50	vendor- K.V. Assoc.
ENGINEERING/DESIGN			8-15% of system cost	
DIFFUSER STACKS				
	CARBON STEEL	4 IN	$8/FT	Add 40% for installation
		6 IN	$10/FT	
	STAINLESS STEEL	4 IN	$30/FT	
		6 IN	$40/FT	

Blowers/Vacuum Pumps

A vacuum pump or a positive displacement blower may be used to provide the power for the SVE system. Fans and pumps are characterized by the flow rate that can be achieved, the horsepower, and the vacuum that can be induced. Vacuum is usually expressed in terms of inches of water (in H_2O) or millimeters of mercury (mm Hg). A large selection of commercial blowers exists and numerous vendors provide these blowers (see Appendix C). The price for this equipment varies with the fan size and power, with a wide variety of combinations available.

Blowers often must be spark and explosion-proof, which will raise the price considerably. Explosion-proof blowers range in price from $1,700 (1 hp) to $6,000 (30 hp). In contrast, blowers that are not so designed range from $300 (1 hp) to $1,900 (10 hp). Prices for the blowers are normally determined by both fan size (6" blades, 8" blades, 10" blades, etc.) and by rating (flow rate). Prices for specific units should be obtained from vendors. Table 12 shows some representative prices.

Monitoring Equipment

Monitoring of the extracted vapor stream is a vital part of SVE design and operation. Monitoring equipment should allow the determination of the vacuum air flow, and vapor characteristics and concentrations.

The vacuum can be measured with a magnehelic gauge. Gauges are typically located at each extraction well and prior to the blower. The cost of magnehelic gauges can range from $50 to $75. Gauges at each well may be substituted by a quick-coupling sampling port. Air flow, expressed in standard cubic feet per minute (scfm) to normalize flow readings taken at different pressures, can be measured in-line by an annubar flowmeter or at flow ports using portable equipment. Air flow should be measured at each well and upstream of the blower. Annubar flow meters cost about $300. Quick-coupling sampling ports with two or three connection can be purchased for $25.

Monitoring of the concentrations and composition of the extracted vapors is critical in determining vapor treatment alternatives and operation procedures. Quantitative vapor concentration can be determined using an organic vapor analyzer (OVA), total hydrocarbon analyzer (THA) or a combustible gas indicator (CGI). Vapor components and concentrations can be determined using a gas chromatograph (GC). The vapor concentration is usually monitored between the demister (or knockout pot) and the blower. In carbon adsorption systems, monitoring may also occur in the exhaust from the carbon bed.

The choice of a specific monitor for vapor concentration will depend on cost, operational range, and sensitivity required. An OVA or an HNu is normally used in conjunction with quick-coupling sampling ports. The OVA uses a flame ionization detector (FID) to determine the total hydrocarbons present. The Foxboro Company offers two portable OVA units. The 108 GC is capable of monitoring VOCs in the range of 0 to 10,000 ppm (0 to 1 percent), while the 128 GC can monitor vapor streams of 0 to 1,000 ppm (0 to 0.1 percent). A 1/10 dilution is available for these OVA units that allows for monitoring of a vapor stream of up to 100,000 ppm (10 percent) with the OVA 108 GC model. Vapor monitoring systems such as the Foxboro OVA models 108 and 128 sell for $6,100 and $7,300 (units have GC capability). The dilutor costs an additional $450.   The HNu, which is also used for monitoring VOCs, is not practical for use with SVE systems because the moisture in the vapor stream, which is present even after moisture reduction from a demister, renders the HNu inoperative.

Total hydrocarbon analyzers can be set in line for continuous sampling. THAs also use an FID for contaminant detection.   J.U.M. Engineering manufactures several types of total hydrocarbon analyzers. For example, the model VE7 has a range of 0 to 100,000 ppm (0 to 10 percent) and has a maximum sensitivity of 1 ppm as methane. These range in price from $15,000 to $21,000. Thermo Environmental Instruments Inc. offers the Model 810 THC analyzer, which is capable of monitoring vapor stream in the range of 0.1 to 10,000 ppm. Other vendors include Byron Instruments and Analytical Instrument Development, Inc.

The combustible gas indicator measures the lower explosive limit (LEL) of the vapor stream. The LEL is the lowest concentration by volume in air at which a combustible gas or vapor will explode, ignite, or burn when there is an ignition source. This equipment is far less sensitive to vapor concentration than the aforementioned FIDs. Limitations of CGI use include:

- dependence on temperature;

- sensitivity is based only on calibration gas; and

- filament may be fouled by leaded gasoline and halogens.

Gastech manufactures a portable CGI (Model 1314). Stationary models are also available. Portable CGIs from Gastech (Model 1314) cost approximately $1,000. Stationary models sell for $1,000 to $2,000.

Gas chromatography equipment is capable of determining the concentrations of specific components. Both portable and laboratory grade units are available. GC units cost from $20,000 to over $30,000.

Sampling probes may be used to monitor soil gas and/or vacuum at locations within the periphery of remediation. Probes are available to monitor vacuum and for vapor sampling. K-V Associates Inc. offers a vapor monitoring probe that can be installed in the vicinity of the SVE system to monitor cleanup. This system is capable of sampling both the soil gas and groundwater and sells for $30. Hollow steel probes are also available, which are driven into the ground for soil gas sampling and vacuum measurements.

Vapor Pretreatment

The extracted vapors are normally pretreated prior to the emission control unit. Air water separators ("knock-out pots") decrease the velocity of the vapor stream and allow water droplets and sediment to fall out via gravity. They can be very simple (e.g., a 55-gallon drum) or may incorporate level controls and other instrumentation. The size depends on the flow rate (to reach a minimum residence time); a typical size range is 800 to 1200 gallons. Construction materials vary but may be cast iron, stainless steel, or similar material. Demisters are often incorporated into the vapor pretreatment process. These screens are capable of removing particles down to microns in size by coalescing these droplets on the demister material.

Duall Industries, Inc. manufactures a variable sized demister ranging in cost from $700 to $1,000 for flow volumes of 100 to 1,000 scfm. Water Resources Associates, Inc. manufactures knockout pots for use with their complete thermal incineration SVE systems. The cost for knockout pots may range from $1500 to $2500 according to size and flow rate capabilities.

Sidestream Treatment

Liquids that accumulate in the air-water separator must be treated on-site, disposed off-site (to a sewer line), or removed by truck. On-site groundwater treatment is often occurring simultaneously with SVE. In such cases, the air-water separator sidestream is added to the groundwater for treatment, typically by air stripping or granular activated carbon (GAC). Where the volume of the sidestream is small compared to the flow of the groundwater treatment method, the marginal cost of treating this sidestream is minimal. Where on-site treatment is desired and groundwater treatment is not occurring, liquid phase GAC can be installed. Carbon is available in small, easily installed units that are appropriate for the flows expected from the vapor pretreatment units. The sidestream can also be removed off-site by tanker trucks.

Emission Control

The vapors removed from the subsurface must normally be treated to reduce the vapor concentration prior to release to the atmosphere, depending on local regulations. Several options are available for vapor treatment, including carbon adsorption, catalytic oxidation, thermal incineration, combination systems, and internal combustion engines. Where vapor treatment is not required, diffuser stacks are used to allow safe dispersal of the extracted vapors. These options are appropriate for different ranges of vapor phase concentration.

The cost of vapor treatment can be a significant portion of the total SVE system cost. Care must be taken to ensure that the most cost-effective vapor treatment option is used. This determination must be based on the vapor discharge standards, the extracted vapor concentration, the expected mass removal over the life of the system, and several other variables. The operations cost of vapor treatment may dominate the calculation of the system cost, especially for GAC treatment systems. For this reason, the forecast for expected removal rate becomes even more important.

Carbon adsorption is a widely used vapor treatment method in industrial and SVE settings. It is applicable to a variety of vapor contaminants and can achieve very high removal rates when required. Carbon is economical only for relatively low mass removal rates; however, when mass removal rates are high, the cost of replacing or regenerating the carbon may be prohibitive.

Carbon adsorption systems are available in a large variety of sizes from numerous vendors; Table 13 shows a partial list of vendors with their respective products. These systems are available in very small systems (55 gallon drums holding less than 200 pounds of carbon) through larger, skid-mounted systems (up to 5,700 pounds of carbon). For very large installations, vendors can customize carbon to the specific requirements of the site. Carbtrol offers the G-1, G-2, G-3, and G-5 canisters that are rated for various air flows. These systems are modified drums containing 200, 170, 140 and 2000 lbs of activated carbon, respectively. The G-1 system, rated at 100 scfm, costs $660, the G-2 (300 scfm) costs $985, the G-3 (500 scfm) costs $1100 and the G-5 (600 scfm) costs $11,000. TIGG Corporation offers a Nixtox series N500 DB, N750 DB and N1500 DB (deep bed) systems that contain 1900, 3200, 5700 pounds of virgin carbon, respectively. Calgon Carbon Corporation offers a large variety of carbon adsorbers. The Ventsorb canister can handle average flows up to 100 cfm or high flows from 400 to 11000 cfm.

TABLE 13. SVE VAPOR TREATMENT UNIT COSTS*

TREATMENT	VENDORS	MODEL	MAXIMUM FLOW (scfm)	CAPITAL COSTS ($)	RENTAL $/month	LEASE PERIOD	OPERATION	EQUIPMENT INCLUDED	NOTES
CARBON ADSORPTION	CARBTROL	SVX	105	18400	1540	1 yr.+ deposit	hp x kw/hr	a skid mounted system including a blower,demister, controls, guages, valves, & flow ammeter	(uses G-1 or G-5 carbon)
			250	21900	1830	1 yr.+ deposit			
			500	30000	1900	1 yr.+ deposit			
	CONTINENTAL RECOVERY SYSTEMS	MANUAL BEDS							
		1 BED		20000	2400		5 hp	includes controls, blower & steam regenerator	
		2 BEDS		26000	2900				
		3 BEDS		32000	3325				
		4 BEDS		38000	3650				
		5 BEDS		44000	3875				
		6 BEDS		50000	4000				
		AUTOMATIC REGENERABLE BEDS		149000	7500	6 mo		fully remote monitored trailer	
	CALGON	HIGH FLOW VENTSORB						skid mounted system includes fan, flexible connector, & damper	
		28 in dia	400	5400			0.5 hp		
		36 in dia	600	8550			3 hp		
		48 in dia	1100	11000			5 hp		
CATALYTIC OXIDATION	DEDERT CORP.	CATOX	1000	85000			fuel	skid mounted system includes reactor, blower, heat exchanger, boiler, & water heater	
			5000	200000					
	CSM SYSTEMS	MODEL 2B TORVEX	200	37500				skid mounted system includes burner, catalyst,control panel, & blower.	treats chlorinated solvents air dillution system=$20,000 trailer = $8500
		MODEL 5A+heat ex.	500	60000					
		MODEL 5B TORVEX	500	50000					
		MODEL 10B TORVEX	1000	70000					
	ORS	CATALYTIC SCAVENGER						skid mounted includes installation heater, control module, heat exchanger, & catalyst	no chlorenated solvents catalyst replacement = $2800
		1282001	200	63000	6600/11584	12mo/6mo	20kw		
		1282002	500	78000			35kw		
		1282008	300	90000	15000 9900/17377	1 mo 12mo/6mo	20kw	additionally includes (2) 5hp blowers,(2)catalysts, piping, filters, guages, valves, flame arrestor, enclosure & trailer	

(continued)

TABLE 13. SVE VAPOR TREATMENT UNIT COSTS*
(continued)

TREATMENT	VENDORS	MODEL	MAXIMUM FLOW (scfm)	CAPITAL COSTS ($)	RENTAL $/month	LEASE PERIOD	OPERATION	EQUIPMENT INCLUDED	NOTES
	WATER RESOURCES ASSOCIATES	AB15-5-SVS	100	11200					No warranty or process efficiency is extended when flow rates are in excess of design capabilities of not more than 12,000 ppm total hydrocarbons
		AB19-10-SVS	210	15300					
		AB22-10-SVS	320	18400					
		AB24-10-SVS	420	20100					
		AB22-15-SVS	570	22900					
		AB15-5-SVS	100	23000	3850		fuel	Includes burner, blower, flame arrestor,guages, valves, filters, knockout pot, & sampling port skid mounted w/enclosure, fence & control panel	
		AB19-10-SVS	210	28000	4675		1.5 hp		
		AB22-10-SVS	320	32000	5350				
		AB24-10-SVS	420	36000	6000				
		AB22-15-SVS	570	40000	6675				
COMBINATION SYSTEMS	HASSTECH	MMC-5	100	60000			fuel	includes thermal incinerator, catalytic oxidizer, vacuum pump compressor, control &analytical instruments, valves, gauges,& fuel system	
		MMC-2 w/trailer	30						
		MMC-3	1000						
	CEMI	RANGER	200	55000			fuel maint. contract $7600/yr	skid mounted includes installation (2)carbon beds, heat exchanger, thermal incinerator,controls, blower, valves, piping,VOC sensor, regeneration steam blower	
CARBON CANNISTERS	CARBTROL	G-SERIES cannisters							carbon can be reactivated at 3-13% discount from origina purchase price
		G-1 200 lb	100	650					
		G-2 170 lb	300				: :		
		G-3 140 lb	500						
		G-5 2000 lb	600	11000					
	CALGON	VAPOR-PAC 1800 lb	1000	5600					
		VENTSORB cannister	100	764			: :		
		HIGH FLOW VENTSORB cannisters 28 in	400	1700					
		36 in	600	4000					
		48 in	1100	6400					
	TIGG	NUXTOX SERIES							Deep bed units
		N5000B	500	7050			: :		
		N750DB	750	11850					
		N1500DB	1500	19150					
		BOXSORBER 6x6	2200	13750					
		BOXSORBER 8x8	4000	20500					

*1989 Estimates. Contact vendor for actual prices.

The high flow model is also available as a skid-mounted unit that includes a fan, flexible connectors, and a damper. The canisters range in price from $760 to $6,330, while the skid-mounted models cost from $5,400 to $10,700.

The carbon may be virgin (unused) or reactivated. Purchase of reactivated carbon usually saves three to thirteen percent off the price of virgin carbon. For example, the virgin G-1 (200 lb) canisters offered by Carbtrol sell for $660, while a reactivated canister sells for $640. Larger containers are usually charged on a weight basis. Environtrol reactivates carbon for $1.15 lb plus transportation costs. A one time RCRA Toxic Characteristics Leaching Procedure (TCLP) test is required ($2800 to $3000) for hazardous materials.

An alternative to the replacement of canisters and off-site reactivation is a recycling carbon system. Such systems regenerate the carbon in place, usually using steam to desorb the contaminants. The contaminant/steam mixture is then drawn off and treated or disposed. Continental Recovery System Inc. offers this type of system. The system comes in several sizes using from one to six carbon beds.

Manually-operated systems cost from $20,000 (one bed) to $50,000 (six beds). A fully automated, remotely-monitored, trailer-mounted system costs $150,000 or leases for $7,400/month on a 6-month lease. The cost effectiveness of the system depends on the mass removal rate (Figure 27). The system is initially more costly than non-regenerative systems, but reduced carbon usage may make it a cheaper option on a life-cycle basis.

Use of carbon for vapor treatment may necessitate the need for a heat exchanging unit to cool extracted vapors because temperatures increase due to the heat of compression from the blower. This will ensure maximum contaminant uptake. Alternatively, the GAC can be placed prior to the blower in the treatment train.

Thermal incineration of contaminant vapors is often an excellent treatment option when vapor concentrations are high. Vapors are combusted at temperatures of 1000 to 1400 degrees F or higher, leading to destruction of over 95 percent of the influent contaminant concentration.

Supplemental fuel may be required to maintain the requisite temperatures for adequate removal. The amount of supplementary fuel depends on the vapor concentration; some vendors report that at gasoline concentration above 12,000 ppm, the flame is self-sustaining; at concentrations below this figure, increasing amounts of fuel are needed in proportion to the contaminant concentration. The operating cost of a thermal incineration system is obviously greatly affected by the amount of supplementary fuel required. Propane, which costs about $1.00/gallon, is often used for this purpose.

While higher contaminant concentrations make this method cheaper, safety concerns increase with higher concentrations. Highly volatile contaminants (such as gasoline) become explosive in certain concentration ranges (Figure 25). This range is limited by the lower explosive limit (LEL) and the upper explosive limit (UEL). When the extracted vapors are at very

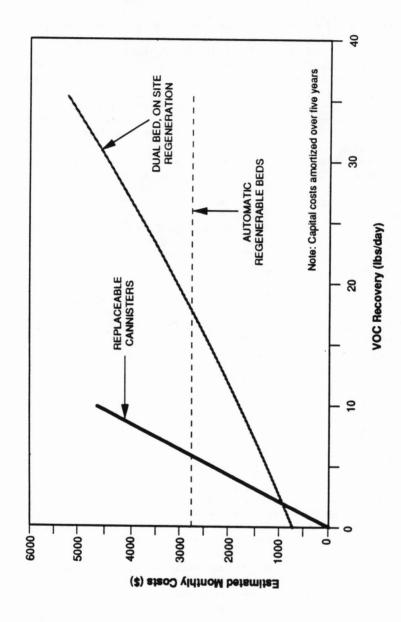

Figure 27. Activated Carbon Systems Cost Comparison

high concentrations, fresh air must be mixed with the vapors to reduce the concentration to a safe level.

Table 13 shows the cost for various incineration units. These units are available in prepackaged units that include the burner unit, blowers, sampling valves, and other appurtenances. Capital costs depend on the flow rate to be treated, and range from $23,000 (for 100 scfm) to $40,000 (570 scfm) from one vendor. A smaller unit (70 scfm) is available for $12,000. A heat recovery system, which uses the exhaust to preheat the incoming vapors, can result in substantial energy and cost savings.

Catalytic oxidation systems employ a catalyst to facilitate the oxidation of contaminants, and thus operate at much lower temperatures (600 to 800 degrees F) than direct thermal incineration while achieving destruction and removal efficiencies (DREs) above 95 percent. The catalyst is a precious metal formulation (typically platinum or palladium) and can be in the form of either beads or a honeycomb bed. Like thermal incineration, this method traditionally has not been appropriate for chlorinated organics due to the potential for forming chlorine gas. The new catalyst described by Trowbridge and Malot (1990), however, may allow catalytic oxidation of vapor streams containing chlroinated solvents.

Catalytic oxidation requires careful monitoring to prevent overheating of the catalyst, resulting in its destruction. If the concentration of vapors in the extracted air is above about 3000 ppm, the vapor stream must be diluted with fresh air to remain below this cutoff level. At low concentrations supplemental fuel (propane) may be needed to maintain the required temperatures. Safety is also a concern for catalytic oxidation. This method is best suited for concentrations below ten percent of the LEL.

Catalytic oxidation units are available that can handle flows as little as 30 to 40 scfm to more than 50,000 scfm for large, permanent industrial facilities. Hasstech offers a trailer mounted unit (MMC-2) that can handle 30 to 40 scfm. ORS offers the Catalytic Scavenger in a 20 kw model (200 scfm) and 35 kw model (500 scfm) that sell for $60,000 and $75,000, respectively. Installation and training will cost $3000 for these units. CSM Systems, Inc. produces the Torvex series Model 5A, 5B (500 scfm) and Model 10B (1000 scfm) that sell for $50,000 and $70,000, respectively. A trailer ($8,500) and ADS dilution system ($20,000) are available for these models. Larger catalytic oxidation systems are also available from CSM and Dedert Corporation. Dedert sells "field ready" units rated at 5000 scfm for $200,000.

Combination vapor treatment systems that combine different treatment options are available from some vendors. These systems often have higher capital costs than simpler systems but may be more economical on a life-cycle basis, depending on the mass removal rate and other considerations. Combination systems often provide some means to destroy (incinerate) the contaminants. This is an important consideration and reduces the exposure to liability arising from transporting contaminants off-site.

Hasstech Combination Systems, Inc. offers a trailer mounted system that incorporates both catalytic oxidation and thermal incineration. This system can accommodate the efficiency of thermal incineration at high contaminant concentration and allows for the use of catalytic oxidation as the concentration in the vapor stream decreases over time. Hasstech's hybrid system (MMC-5) is sold for $60,000. Supplemental fuel (propane or natural gas) is required for operation.

CEM Inc. (CEMI) markets a unit that uses both carbon adsorption and thermal incineration. VOCs are adsorbed onto the carbon, which is then regenerated by steam. The VOC-laden stream is then oxidized by thermal incineration. This system allows for on-site elimination of VOCs and reduces the fuel costs often associated with thermal incineration. This system operates at a flow from 40 to 500 scfm and sells for $50,000, and requires either natural gas or propane for start up and heating of the water to strip the carbon. No additional fuel is required to combust the contaminants in most cases.

Internal combustion engines have also been used to destroy vapor phase concentrations. Several vendors sell or rent such units. RSI Inc. sells a unit powered by a Ford, 4-cylinder, 63 hp, 140 cubic inch capacity engine capable of accepting 30 to 60 scfm. The engine itself sells for $56,500, which includes 8 hours of a technician's time during setup. RSI also leases this engine for $4,800 per month with a six-month minimum. RENMAR sells a Ford 6-cylinder, 100 hp, 300 cubic inch engine with a design flow rate of 100 scfm. This engine sells for $45,000 or rents for $4,000/month. Wayne Perry Construction rents their system, which is powered by a 460 cubic inch Ford engine reportedly capable of passing 400 scfm, for $7,500/month. This price also includes some consulting services such as geological services.

Diffuser stacks are usually constructed from either carbon steel or stainless steel. They merely direct vapors into the atmosphere. This system is simple and inexpensive, but only an option where treatment of the vapors is not required. Diffuser stacks should be designed to minimize health risks. The cost of diffuser stacks depends on the height required and the material of construction.

Other Costs

Implementation of an SVE system will entail other costs that are not strictly capital costs or O&M costs, and include system design, engineering, permit acquisition, contingencies and other miscellaneous costs. These costs are often treated as capital costs. Engineering and design fees are often approximated by 10 to 15 percent of the system cost, as are contingencies. These and other costs are highly site-specific, however, and the figures quoted here are arbitrary.

OPERATION AND MONITORING COSTS

The operation and monitoring (O&M) costs may comprise a significant portion of the overall remediation cost. These costs are composed mainly of power requirement for the blowers and condenser (if necessary); vapor treatment, including fuel costs for incineration methods and GAC regeneration/replacement; monitoring and analyses for SVE progress and cleanup attainment determination; and other on-going costs such as labor, which depends most highly on whether the system is operated manually or with a microprocessor. These costs are discussed separately below.

Power Requirements

The cost for electric power depends on the horsepower of the fan or blower, the hours of operation, and the local cost of electricity. The formula for determining the cost is: (0.75) x (fan horsepower) x (electricity cost in $/kw-hr) x (hours of operation). For example, if a 10-hp blower was operated continuously and electricity averaged $0.10/kw-hr, the daily cost for power would be 10 x 0.75 x $0.10 x 24 - $18.00 per day. Pulsed operation -- operating the blowers intermittently -- would save power costs by decreasing the hours of operation. Power may also be required if heat exchangers are used.

Vapor Treatment

The operating cost of the vapor treatment depends on the treatment method, the concentration of contaminants, and the flow rate. Generally, the cost for GAC adsorption increases while the cost for incineration and oxidation decreases with higher vapor concentrations. GAC treatment costs will be dominated by carbon replacement and regeneration; incineration and oxidation treatment will be dominated by fuel costs to sustain incineration.

Carbon adsorption. Adsorption of contaminants from the vapor phase concentrates the contaminants onto the carbon. When the carbon's capacity to hold contaminants is used up, the carbon is considered "spent" and must be replaced or regenerated. Obviously, higher mass removal rates (flow rate x concentration) will result in more frequent carbon replacement and higher costs.

Carbon costs vary depending on the type and quantity ordered, and may range up to $2.00/lb. Regenerated carbon costs 87 to 97 percent of virgin carbon cost. One vendor quoted $1.15/pound as the cost of regenerated carbon for large orders. Table 13 shows costs for virgin carbon units. Hinchee et al. (1987) state that, as a rule of thumb, carbon costs about $20 per pound ($130 per gallon) of gasoline removed.

Where carbon use is desired and mass removal rates are high, on-site regeneration may become economical. Continental Recovery Systems offers a unit that uses steam to regenerate carbon in place. Other vendors offer units that regenerate the carbon and then incinerate the contaminants. These combination units are more costly initially but save on O&M costs; Figure 27 shows the mass removal rates when the system are cheaper on a life-cycle cost basis. The determination of the most cost-effective option is site-specific and is normally found through pilot system results.

Thermal incineration. Thermal incineration requires supplementary fuel for vapor concentrations below about 12,000 ppm. This fuel is typically propane or LPG, which cost about $1.00/gallon. When the BTU value of the contaminants is not sufficient to sustain the required temperature (about 1400 to 1600 degrees F), fuel must be added to maintain proper temperatures.

Catalytic oxidation. This method requires much lower temperatures (600 to 800 degrees F) than thermal incineration and therefore is less costly to operate. Optimal vapor phase concentration for catalytic oxidation is about 3000 ppm; higher concentrations require dilution (to protect the catalyst from destruction) while lower concentration may require supplemental fuel. ORS states that a 200 scfm Catalytic Scavenger costs about $800 per month to operate with no incoming hydrocarbons (i.e., just air); as the hydrocarbon concentration increases, the supplemental fuel requirements decrease.

Internal combustion engines. Like other incineration techniques discussed above, internal combustion engines require supplemental fuel if the vapor phase concentrations are too low to sustain the proper air-fuel ratio. Supplemental fuel is propane or, perhaps, natural gas. RSI reports that 20 gallons of propane per day is typical, with the range from zero to an upper end of 35 gallons. Propane costs about $1.00/gallon. RSI reports that if natural gas is used, the equivalent cost would be about $0.75/gallon.

Monitoring and Analyses

Laboratory sampling for soil, groundwater, and vapor contaminant concentrations is relatively costly. It is therefore judicious to determine carefully the appropriate samples to be collected and the proper methodology for sample collection and preparation. Soil sample analyses will generally cost $75-125 for TPH, $1,100-1,600 for TCLP, $200-250 for VOCs, $100 for BTEX, $450-550 for ABNs, $100-200 for a petroleum "fingerprint" identification, and $70 for routine soil parameters, which include organic carbon and particle size distribution.

Analysis for groundwater sampling costs $100 (TPH), $225 (VOCs), $100 (BTEX), $400-500 (ABNs), and $50 for general groundwater quality parameters. Soil gas analysis using a GC for determination of total hydrocarbons and specific contaminants may cost as much as $250 if sent to a laboratory.

These price ranges reflect the variation among different laboratories and should be used as a guide. Large orders or preferred customers may qualify for discounts. Priority orders that require short turnaround times are surcharged; analyses can often be performed on a 48-hour turnaround, but at double the cost.

References

A.D. Little, Inc. 1987. The Installation Restoration Program Toxicology Guide, Vol. 3. Prepared for the Aerospace Medical Division, Air Force Systems Command, Wright-Patterson AFB, OH.

Alexander, M. 1977. Introduction to Soil Microbiology. 2nd Edition, John Wiley and Sons, New York, NY.

Anastos, G.J., P.J. Marks, M.H. Corbin, and M.F. Coia. 1985. Task 11. In-situ Air Stripping of Soils, Pilot Study, Final Report No. AMXTH-TE-TR-85026. U.S. Army Toxic & Hazardous Material Agency. Aberdeen Proving Grounds.

Applegate, J., J.K. Gentry, and J.J. Malot. 1987. Vacuum Extraction of Hydrocarbons from Subsurface Soils at a Gasoline Contamination Site. Proceedings of the 8th National Conference of Uncontrolled Hazardous Wastes (Superfund '87). Washington, D.C. Nov. 16-18. pp. 273-278.

Aurelius, M.W. and K.W. Brown. 1987. Fate of Spilled Xylene as Influenced by Soil Moisture Content. Water, Air and Soil Pollution. 36:23-31

Bear, J. 1979. Hydraulics of Groundwater. McGraw-Hill, New York.

Bell, C.E., P.T. Kostecki and E.J. Calabrese. 1989. State of Research and Regulatory Approach of State Agencies for Cleanup of Petroleum Contaminated Soils. In: Petroleum Contaminated Soils, Vol. 2. Calabrese, E.J., P.T. Kostecki, and C.E. Bell (Eds.). Chelsea, MI: Lewis Publishers.

Bennedsen, M.B. 1987. Vacuum VOCs from Soil. Pollution Engineering. 19(2):66-69.

Bennedsen, M.B., J.P. Scott, and J.D. Hartley. 1985. Use of Vapor Extraction Sytems for In-situ Removal of Volatile Organic Compounds from Soil. Proceedings of the 5th National Conference on Hazardous Wastes and Hazardous Materials, HMCRI. pp. 92-95.

Benson, R.C. 1988. Surface and Downhole Geophysical Techniques for Hazardous Waste Site Investigations. Hazardous Materials Control. March/April 1(2):8.

Boardman, C.R. and J.W. Strove. 1966. Distribution in Fracture Permeability of a Granitic Rock Mass Following a Contained Nuclear Explosion. J. Petroleum Technology. 18(5):619-623.

Bradford, M., B.J. Marks, and M. Singh. 1989. Comparison of Soil-Gas, Soil and Groundwater Contaminant Levels of Benzene and Toluene. Proceedings of the 9th National Conference on Hazardous Wastes and Hazardous Materials. New Orleans, LA. HMCRI.

Brown, R. and C. Harper. 1989. Above Ground, Bioaugmented Soil Venting. Presented at the Soil Vapor Extraction Technology Workshop, Office of Research and Development, Edison, NJ. June 28-29.

Calabrese, E.J., P.T. Kostecki, and C.E. Bell (Eds.) 1989. Petroleum Contaminated Soils. Vol. 2. Chelsea, MI: Lewis Publishers.

Calgon Carbon Corporation. Undated. Company literature.

Camp Dresser & McKee Inc. 1986. Interim Report, Fate and Transport of Subsurface Leaking from Underground Storage Tanks. Volume I. Prepared for Office of Underground Storage Tanks, USEPA, Washington, DC.

Camp Dresser & McKee Inc. 1988. Interim Report for Field Evaluation of Terra Vac Corrective Action Technology at a Florida LUST Site. Contract No. 68-03-3409. Office of Research and Development, USEPA. Edison, NJ.

Carlson, D.A. and C.P. Leiser. 1966. Soil Beds for the Control of Sewage Odors. J. Water Pollution Control Federation. 38(5):829-840.

Connor, R.J. 1988. Case Study on Soil Venting. Pollution Engineering. 20(8):74-78.

Corey, A.T. 1957. Measurement of Water and Air Permeability in Unsaturated Soils. Soil Sci. Soc. Am. Proc. 21(1):7-10

Crow, W.L., E.P. Anderson, and E.M. Minugh. 1987. Subsurface Venting of Vapors Emanating from Hydrocarbon Product on Groundwater. Ground Water Monitoring Review. 7(1):51-57.

CSM Systems, Inc. 1989. Company literature. Brooklyn, New York.

Danko, J. 1989. Applicability and Limitations of Soil Vapor Extraction. Presented at the Soil Vapor Extraction Technology Workshop, Office of Research and Development, Edison, NJ. June 28-29.

Daugherty, S.J. 1989. The California Leaking Underground Fuel Tank Field Manual: A Guidance Document for Assessment of Underground Fuel Leaks. In: Calabrese, E.J., P.T. Kostecki, and C.E. Bell (Eds.) Petroleum Contaminated Soils, Vol. 2, Chelsea, MI: Lewis Publishers.

Davies, S.H. 1989. The Influence of Soil Characteristics on the Sorption of Organic Vapors. Presented at the Workshop on Soil Vacuum Extraction, R.S. Kerr Environmental Research Laboratory. Ada, OK, April 27-28.

DePaoli, D.W., S.E. Herbes, and M.G. Elliott. 1989. Performance of In-situ Soil Venting System at Jet Fuel Spill Site. Presented at the Soil Vapor

Extraction Technology Workshop, Office of Research and Development, Edison, NJ. June 28-29.

Dev, H. and D. Downey. 1988. Zapping Hazwastes. Civil Engineering. 58(8):43-46.

Donohue, D.A.T. and T. Ertekin. 1982. Gaswell Testing Theory, Practice and Regulation. International Human Resources Development Corporation, Boston, MA. 214 pp.

Dragun, J. 1988. The Soil Chemistry of Hazardous Materials. Hazardous Materials Control Research Institute, Silver Spring, MD.

Dragun, J. and J. Barkach. 1989. Three Common Misconceptions Concerning the Fate and Cleanup of Petroleum Products in Soil and Groundwater. In: Petroleum Contaminated Soil. Volume. 2. Calabrese, E.J., P.T. Kostecki, and C.E. Bell (Eds.). Chelsea, MI:Lewis Publishers.

Duchaine, R.P. 1986. Look Before You Leap: The Case for Restraint from Active Site Remediation. Proceedings of Petroleum Hydrocarbons and Organic Chemicals in Groundwater: Prevention, Detection, and Restoration. Houston, TX.

Elrick, D.E., W.D. Reynolds, and K.A. Tan. 1989. Hydraulic Conductivity Measurements in the Unsaturated Zone Using Improved Well Analyses. Ground Water Monitoring Review. 9:184-193.

EPA. 1986a. Control Technologies for Hazardous Air Pollutants. EPA/625/6-86/014.

EPA. 1986b. Hazardous Waste Management System; Identification and Listing of Hazardous Waste; Notification Requirements; Reportable Quantity Adjustments; Proposed Rule. Federal Register 51, No. 114, 13 June 1986, 21648-21693.

EPA. 1987a. A Compendium of Superfund Field Operations Methods. EPA/540/87/001.

EPA. 1987b. Data Quality Objectives for Remedial Response Activities. EPA/540/G-87/003&004.

EPA. 1988a. "Underground Storage Tanks; Technical Requirements." Federal Register 53, No. 185, 23 September 1988, 37082-37212.

EPA. 1988b. Cleanup of Releases from Petroleum USTs: Selected Technologies. EPA/530/UST-88/001.

EPA. 1988c. Guidance for Conducting Remedial Investigations and Feasibility Studies Under CERCLA, Interim Final OSWER Directive. 9355.3-01.

EPA. 1988d. Survey of State Programs Pertaining to Contaminated Soils. Office of Underground Storage Tanks, March 22.

EPA. 1990. Assessing UST Corrective Action Technologies: Site Assessment and Selection of Unsaturated Zone Treatment Technologies. EPA/600/2-90/011.

Evans, D.D. and D. Kirkham., 1949. Measurement of Air Permeability of Soil In Situ. Soil Science Society America Proceedings. 14:65-73.

Fall, E.W. and W.E. Pickens. 1988. In-situ Hydrocarbon Extraction: A Case Study. Proceedings of the FOCUS Conference on Southwestern Ground Water Issues, Albuquerque, NM.

Farmer, W.J., M.S. Yang, J. Letey, and W.F. Spencer. 1980. Land Disposal of Hexachlorobenzene Wastes: Controlling Vapor Movement in Soil. EPA/600/2-80/119.

Fredericks, R. of In-Process Technology (Sunnyvale, CA). 1990. Personal communication with J. Curtis of Camp Dresser & McKee (Boston, MA) on January 19.

Greene, T. of M-P Pneumatics (Lexington, MA). 1990. Presentation at Camp Dresser & McKee Inc. (Boston, MA). March 21.

Grover, B.L. 1955. Simplified Air Permeameters for Soil in Place. Soil Science Society America Proceedings. 19:414-418.

Hardison, L.C. and E.J. Dowd. 1977. Emission Control Via Fluidized Bed Oxidation. Chemical Engineering Progress. August. pp. 31-35.

Hinchee, R.E., D.C. Downey, and E.J. Coleman. 1987. Enhanced Bioreclamation, Soil Venting and Ground-Water Extraction: A Cost-Effectiveness and Feasibility Comparison. Proceedings of Petroleum Hydrocarbons and Organic Chemicals in Ground Water: Prevention, Detection, and Restoration. Houston, TX.

Hinchee, R.E. 1989. Enhanced Biodegradation Through Soil Venting. Presented at Workshop on Soil Vacuum Extraction, R.S. Kerr Environmental Research Laboratory. Ada, OK. April 27-28.

Hinchee, R.E., D. Downey, and R. DuPont. 1989. Biodegradation Associated with Soil Venting. Presented at the Soil Vapor Extraction Technology Workshop, Office of Research and Development, Edison, NJ. June 28-29.

Hoag, G.E. 1989. Soil Vapor Extraction Research Developments. Presented at the Soil Vapor Extraction Technology Workshop, Office of Research and Development, Edison, NJ. June 28-29.

Hoag, G.E. and M.L. Marley. 1986. Gasoline Residual Saturation in Unsaturated Uniform Aquifer Materials. Journal of Environmental Engineering. 112(3):586-604.

Houston, S.L., D.K. Kreamer, and R. Marwig. 1989. A Batch-Type Testing Method for Determination of Adsorption of Gaseous Compounds on Partially Saturated Soils. Journal of Testing and Evaluation, ASTM. 12(1):3-10.

Howe, G.B., M.E. Mullins, and T.N. Rogers. 1986. Evaluation and Prediction of Henry's Law Constants and Aqueous Solubilities for Solvents and Hydrocarbon Fuel Components Volume I: Technical Discussion. USAFESC Report No. ESL-86-66. U.S. Air Force Engineering and Services Center, Tyndall AFB, FL. 86 pp.

Hunt, J.R., N. Sitar, and K.S. Udell. 1986. Organic Solvents and Petroleum Hydrocarbons in the Subsurface: Transport and Cleanup. University of California at Berkeley, Sanitary Engineering and Environmental Health Research Laboratory. Report 86-11.

Hutzler, N.J., B.E. Murphy, and J.S. Gierke. 1989a. Review of Soil Vapor Extraction System Technology. Presented at the Soil Vapor Extraction Technology Workshop, Office of Research and Development, Edison, NJ. June 28-29.

Hutzler, N.J., J.S. Gierke, and L.C. Krause. 1989b. Movement of Volatile Organic Chemicals in Soils. In: Reaction and Movement of Organic Chemicals in Soils. B.L. Sawhney and K. Brown. Soil Science Society of America, Spec. Publication No. 22. pp. 373-404.

Ikoku, C.U. 1984. Natural Gas Reservoir Engineering. J. Wiley & Sons, New York.

Johnson, J.J. and R.J. Sterrett. 1988. Analysis of In-situ Soil Air Stripping Data. Proceedings of the 8th National Conference on Hazardous Wastes and Hazardous Materials, HMCRI. Las Vegas, NV. April 19-21.

Johnson, M.D. 1989. Soil Bed Treatment. Presented at the Soil Vapor Extraction Technology Workshop, Office of Research and Development, Edison, NJ. June 28-29.

Johnson, P.C., M.W. Kemblowski, and J.J. Colthart. 1990. Quantitative Analysis for the Cleanup of Hydrocarbon-Contaminated Soils by In Situ Soil Venting. Journal of Ground Water. 28(3):413-429.

Johnson, P.C., M.W. Kemblowski, J.D. Colthart, D.L. Byers, and C.C. Stanley. 1989. A Practical Approach to the Design, Operation, and Monitoring of In-situ Soil Venting Systems. Presented at the Soil Vapor Extraction Technology Workshop, Office of Research and Development, Edison, NJ. June 28-29. [Also published (1990): Ground Water Monitoring Review, 10(2):150-178].

Johnson, R.L. 1989. Soil Vacuum Extraction: Laboratory and Physical Model Studies. Presented at the Workshop on Soil Vacuum Extraction, R.S. Kerr Environmental Research Laboratory, Ada, OK. April 28-29.

Jury, W.A., A.M. Winer, W.F. Spencer, and D.D. Focht. 1987. Transport and Transformations of Organic Chemicals in the Soil-Air-Water Ecosystem. Reviews of Environmental Contamination and Toxicology. 99:120-164.

Jury, W.A. and M. Ghodrati. 1989. Overview of Organic Chemical Environmental Fate and Transport Modeling Approaches. In: Reactions and Movement of Organic Chemicals in Soils. B.L. Sawhney and K. Brown. Soil Science Society of America, Special Publication No. 22. Madison, WI. pp. 271-304.

Jury, W.A. and R.L. Valentine. 1986. Transport Mechanisms and Loss Pathways for Chemicals in Soil. In: Vadose Zone Modeling of Organic Pollutants. S.C. Hern and S.M. Melancon (eds.). Chelsea, MI: Lewis Publishers. pp. 37-55.

Kane, M. (Ed.) 1987. Manual of Sampling and Analytical Methods for Petroleum Hydrocarbons in Groundwater and Soil. American Petroleum Institute. Publication No. 4449. Washington, D.C.

Karably, L.S. and K.B. Babcock. 1987. Effects of Environmental Variables on Soil Gas Surveys. Proceedings of the 7th National Conference of Management of Uncontrolled Hazardous Wastes (Superfund '87). Washington, D.C. Nov. 16-18. pp. 97-100.

Keech, D.A. 1989. Surface Venting Research and Venting Manual by the American Petroleum Research Institute. Presented at the Workshop on Soil Vacuum Extraction, R.S. Kerr Environmental Research Laboratory. Ada, OK. April 27-28.

Kerfoot, H.B. 1989. Soil Gas Surveys in Support of Design of Vapor Extraction Systems. Presented at the Soil Vapor Extraction Technology Workshop, Office of Research and Development, Edison, NJ. June 28-29.

Kemblowski, M.W. 1989. Application of Soil Vacuum Extraction - Screening Models and Field Considerations. Presented at the Workshop on Soil Vacuum Extraction. R.S. Kerr Environmental Research Laboratory. Ada, OK. April 27-28.

Kopp, J.F. and G.D. McKee. 1979. Methods for Chemical Analysis of Water and Wastes. USEPA EPA 600-4-79-020, NTIS PB 297686As.

Krauss, E.V., J.G. Oster, and K.O. Thomsen. 1986. Processes Affecting the Interpretation of Trichloroethylene Data from Soil Gas Analysis. Proceedings of the 6th National Conference on Management of Uncontrolled Hazardous Waste Sites. Washington, DC, December 1-3. pp. 138-142.

Krishnayya, A.V., M.S. O'Connor, J.G. Agar and R.D. King. 1988. Vapor Extraction Systems Factors Affecting their Design and Performance. Proceedings of Petroleum Hydrocarbons and Organic Chemicals in Groundwater Prevention, Detection, and Restoration. November 9-11. pp. 547-569.

Levy, B.S. and P.F. Germann. 1988.   Kinematic Wave Approximation to Solute
     Transport Along Preferred Flow Paths in Soil.   Journal of Contaminant
     Hydrology. 3:263-276.

Lighty, J.S., G.D. Silcox, D.W. Pershing, and V.A. Cundy. 1988.   On the
     Fundamentals of Thermal Treatment for the Cleanup of Contaminated Soils.
     APCA 81st Annual Meeting, Dallas, TX.   June 19-24.

Lord, A.E., Jr., R.M. Koerner, V.P. Murphy, and J.E. Brugger. 1987.   In-situ
     Vacuum Assisted Steam Stripping of Contaminants from Soil.   Proceedings
     of the 7th National Conference of Management of Uncontrolled Hazardous
     Wastes (Superfund '87).   HMCRI.   pp.   390-395.

Malot, J.J. 1989. Vacuum: Defense for L.U.S.T. Presented at the Soil Vapor
     Extraction Technology Workshop, Office of Research and Development,
     Edison, NJ. June 28-29.

Malot, J.J. and P.R. Wood. 1985.   Low Cost, Site Specific, Total Approach to
     Decontamination.   Conference on Environmental and Public Health Effects
     of Soils Contaminated with Petroleum Products. University of
     Massachusetts. Amherst, MA.   October 30-31.

Marley, M.C., S.D. Richter, B.L. Cliff, and P.E. Nangeroni. 1989. Design of
     Soil Vapor Extraction Systems - A Scientific Approach. Presented at the
     Soil Vapor Extraction Technology Workshop, Office of Research and
     Development, Edison, NJ. June 28-29.

Marrin, D.L. 1988.   Soil Gas Sampling and Misinterpretation.   Ground Water
     Monitoring Review. 8(2):51-54.

Michaels, P.A. and M.K. Stinson. 1989. Technology Evaluation Report.   SITE
     Program Demonstration Test, Terra Vac In-situ Vacuum Extraction System,
     Groveland, MA. Contract No. 68-03-3255.

Millican, R. of South Coast Air Quality Management District (Los Angeles, CA).
     1989.   Personal communication with J. Curtis of Camp Dresser & McKee
     (Boston, MA) on December 4.

Mishra, S., J.C. Parker, and N. Singhal. 1989.   Estimation of Soil Hydraulic
     Properties and Their Uncertainty from Particle Size Distribution Data.
     Journal of Hydrology. 108:1-18.

Munz, C. and P.V. Roberts. 1987. Air-Water Phase Equilibria of Volatile
     Organic Solutes. JAWWA. 79(5):62-69.

Mutch, R.D., Jr., A.N. Clarke, D.J. Wilson, and P.D. Mutch. 1989.   In-situ
     Vapor Stripping Research Project: A Progress Report.   Presented at the
     Soil Vapor Extraction Technology Workshop, Office of Research and
     Development, Edison, NJ. June 28-29.

Neff, C.R. 1989. Biofilter Technology for Odor Control.   BioFiltration, Inc.
     (Company publication), Gainsville, FL.

Oster, C.C. and N.C. Wenck. 1988. Vacuum Extraction of Volatile Organics from Soils. Proceedings of the 1988 Joint CSCE-ASCE National Conference on Environmental Engineering, Vancouver, B.C., Canada. pp. 809-817. July 13-15.

Payne, F.C. and J.B. Lisiecki. 1988. Enhanced Volatilization for Removal of Hazardous Waste from Soil. Proceedings of the 8th National Conference on Hazardous Wastes and Hazardous Materials, HMCRI. Las Vegas, NV. April 19-21.

Perry, W. of Wayne Perry, Inc. (Buena Park, CA). 1989. Personal communication with J. Curtis of Camp Dresser & McKee (Boston, MA) on December 1.

Pitchford, A.M., A.T. Mazzella, and K.R. Scarbrough. 1988. Soil-Gas and Geophysical Techniques for Detection of Subsurface Organic Contamination. EPA/600/S4-88/019.

Potter, P.L. 1989. Analysis of Petroleum Contaminated Soil and Water: An Overview. In: Petroleum Contaminated Soils, Vol. 2. Calabrese, E.J., P.T. Kostecki, and C.E. Bell (Eds.). Chelsea, MI:Lewis Publishers.

Prokop, W.H. and H.L. Bohn. 1985. Soil Bed System for Control of Rendering Plant Odors. J. Air Pollution Control Association. 35(12):1332-1338.

Regalbuto, D.P., J.A. Barrera, and J.B. Lisiecki. 1988. In-situ Removal of VOCs by Means of Enhanced Volatilization. Proceedings of the Conference on Petroleum Hydrocarbons and Organic Chemicals in Ground Water: Prevention, Detection and Restoration. Houston, TX, November 9-11.

Reible, D.D. 1989. Introduction to Physico-Chemical Processes Influencing Enhanced Volatilization. Presented at the Workshop on Soil Vacuum Extraction. R.S. Kerr Environmental Research Laboratory, Ada, OK. April 27-28.

Reisinger, H.J., D.R. Burris, L.R. Cessar, and G.D. McCleary. 1987. Factors Affecting the Utility of Soil Vapor Assessment Data. Proceedings of the First National Outdoor Action Conference on Aquifer Restoration, Ground Water Monitoring and Geophysical Methods. Las Vegas, NV, May 18-21. pp. 425-435.

Robbins, G.A., R.D. Bristol, and V.D. Roe. 1989. A Field Screening Method for Gasoline Contamination Using a Polyethylene Bag Sampling System. Ground Water Monitoring Review. 9(4):87-97.

Sabadell, G.P., J.J. Eisenbeis, and D.K. Sunada. 1989. The 3-D Model CSUGAS: A Management Tool for the Design and Operation of Soil Venting Systems. Proceedings of the 9th Annual Conference on Hazardous Waste and Hazardous Material. New Orleans, LA. pp. 177-182.

Silka, L.R. 1988. Simulation of Vapor Transport Through the Unsaturated Zone - Interpretation of Soil Gas Surveys. Ground Water Monitoring Review.

8(2):115-123.

Silka, L.R., H. Cirpili, and D.L. Jordan. 1989. Modeling Applications to Vapor Extraction Systems. Presented at the Soil Vapor Extraction Technology Workshop, Office of Research and Development, Edison, NJ. June 28-29.

Smith, P.G. and S.L. Jensen. 1987. Assessing the Validity of Field Screening of Soil Samples for Preliminary Determination of Hydrocarbon Contamination. Proceedings of the 7th National Conference on Management of Uncontrolled Wastes (Superfund '87). pp. 101-103.

Smith, R.V. 1983. Practical Natural Gas Engineering. PennWell Publishing, Tulsa, OK. 252 pp.

Spittler, T.M. 1980. Use of Portable Organic Vapor Detectors for Hazardous Waste Site Investigations. Presented at the U.S. EPA National Conference on Management of Uncontrolled Hazardous Waste Sites. Washington, D.C. October 15-17.

Stephanatos, B.N. 1988. Modeling the Transport of Gasoline Vapors by an Advective Diffusive Unsaturated Zone Model. Proceedings of the Conference on Petroleum Hydrocarbons and Organic Chemicals in Ground Water: Prevention, Detection, and Restoration. Houston, TX. November 9-11. pp. 591-611.

Stever, D.W. 1989. Emerging Legal Issues Related to Air Toxics. Presented at Environmental Hazards Conference & Exposition, Hartford, CT, October 10-12.

Stinson, M.K. 1989. EPA SITE Demonstration of the Terra Vac In Situ Vacuum Extraction Process in Groveland, Massachusetts. J. Air Pollution Control Association. 39(8):1054-1062.

Stonestrom, D.A. and J. Rubin. 1989. Air Permeability and Trapped-Air Content in Two Soils. Water Resources Research. 25(9):1959-1969.

Tate, R.L. 1987. Soil Organic Matter. John Wiley and Sons.

Thompson, G.M., and D.L. Marrin. 1987. Soil Gas Contaminant Investigations: A Dynamic Approach. Ground Water Monitoring Review. 7(3):88-93.

Trowbridge, B.E. and J.J. Malot. 1990. Soil Remediation and Free Product Removal Using In-Situ Vacuum Extraction with Catalytic Oxidation. Proceedings of the Fourth Annual Outdoor Action Conference on Aquifer Restoration, Ground Water Monitoring and Geophysical Methods. Las Vegas, NV.

U.S. Department of Health and Human Services. 1985. NIOSH Pocket Guide to Chemical Hazards.

Van Groenewoud, H. 1968. Methods and Apparatus for Measuring Air Permeability of the Soil. Soil Science Society America Proceedings.

106(4):275-279.

Weast, R.C. (Ed.) 1981. Handbook of Chemistry and Physics. 61st Edition. Boca Raton, FL: CRC Press.

Weber, B. Undated. In Situ Soil Venting Utilizing M-D Pneumatics Blowers. Company literature, Springfield, MO.

Woodward-Clyde Consultants. Undated. Company literature.

Zenobia, K.E., D.K. Rothenbaum, S.B. Charjee, and E.S. Findlay. 1987. Vapor Extraction of Organic Contaminants from the Vadose Zone - A Case Study. Proceedings of the FOCUS on Northwestern Ground Water Issues. Portland, OR, May 5-7. pp. 625-646.

Bibliography

American Petroleum Institute. 1984. Forced Venting to Remove Gasoline Vapor from a Large-Scale Model Aquifer. American Petroleum Institute. Environmental Affairs Department. Washington, D.C.

American Petroleum Institute. 1985. Subsurface Venting of Hydrocarbon Vapors from an Underground Aquifer. API Publication No. 4410. Health and Environmental Sciences Department. Washington, D.C.

AWARE, Inc. 1987. Phase I - Zone I Soil Decontamination through In-situ Vapor Stripping Processes, Final Report. Small Business Innovative Research Program, U.S. EPA.

Baehr, A.L. and G.E. Hoag. 1985. Air Modeling and Experimental Investigation of Venting Gasoline from Contaminated Soils. University of Massachusetts Conference, Amherst, MA.

Baehr, A.L., G.E. Hoag, and M.C. Marley. 1989. Removing Volatile Contaminants from the Unsaturated Zone by Inducing Advective Air-Phase Transport. Journal of Contaminant Hydrology. 4:1-26.

Batchelder, G.V., W.A. Panzeri, and H.T. Phillips. 1986. Soil Ventilation for the Removal of Adsorbed Liquid Hydrocarbons in the Subsurface. Proceedings of the NWWA/API Conference on Petroleum Hydrocarbons and Organic Chemicals in Groundwater. Houston, TX. November 12-14.

Batchelder, G., A.E. Perry, and K. Brody. 1989. An Integrated Technology Approach to Remediation of Chlorinated Hydrocarbons. Presented at the Soil Vapor Extraction Technology Workshop, Office of Research and Development, Edison, NJ. June 28-29.

Bouchard, D.C. 1989. The Role of Sorption in Contaminant Transport. Presented at the Workshop on Soil Vacuum Extraction. R.S. Kerr Environmental Research Laboratory Ada, OK. April 27-28.

CH$_2$M-Hill, Inc. 1985. Verona Well Field - Thomas Solvents Company, Battle Creek, Michigan, Operable Unit Feasibility Study. U.S. EPA. Chicago, IL.

CH$_2$M-Hill, Inc. 1987. Operable Unit Remedial Action, Soil Vapor Extraction at Thomas Solvents Company, Battle Creek, Michigan, Quality Assurance Project Plan. U.S. EPA. Chicago, IL.

Chiou, C.T. 1989. Theoretical Consideration of the Partition Uptake of Nonionic Organic Compounds by Soil Organic Matter. B.L. Sawhney and K. Brown (5th ed.). In: Reactions and Movement of Organic Chemicals in Soil. Soil Science Society of America, Inc. Madison, WI.

Chiou, C.T., D.E. Kile, and R.L. Malcolm. 1988. Sorption of Vapors of Some Organic Liquids on Soil Humic Acid and Its Relation to Partitioning of Organic Compounds in Soil and Organic Matter. Environmental Science & Technology. 22(3):298-303.

Connor, R.J. 1988. Gasoline Recovery Using In-situ Soil Venting, with Field Recharged Carbon Adsorption - A Case Study. Proceedings of the 6th Annual Hazardous Materials Management Conference International, Atlantic City, NJ.

Curtis, J.T., T.A. Pedersen, and C.Y. Fan. 1990. Soil Vapor Extraction Technology. Presented at the 16th Annual EPA Hazardous Waste Research Symposium. Cincinnati, OH. April 3-5.

DiGuilio, D.C., J.S. Cho, R.R. DuPont, and M.W. Kemblowski. 1990. Conducting Field Tests for Evaluation of Soil Vacuum Extraction Apparatus. Proceedings of Fourth National Outdoor Action C onference on Aquifer Restoration, Ground Water Monitoring, and Geophysical Methods. NWWA, Las Vegas, NV.

Dragun, J. 1989. Recovery Techniques and Treatment Technologies for Petroleum and Petroleum Products in Soil and Groundwater. In: Kostecki, P.J. and E.J. Calabrese (Eds.) Petroleum Contaminated Soils, Volume 1, Chelsea, MI: Lewis Publishers.

Dynamac Corporation. 1986. Literature Review of Forced Air Venting to Remove Subsurface Organic Vapors from Aquifers and Soil. Subtask Statement No. 3. U.S. Air Force Engineering and Services Center. Tyndall AFB, FL.

Ellgas, R.A., and N.D. Marachi. 1988. Vacuum Extraction of Trichloroethylene and Fate Assessment in Soils and Groundwater: Case Study in California. Proceedings of the 1988 Joint CSCE-ASCE National Conference on Environmental Engineering. Vancouver, B.C., Canada. pp. 794-801.

EPA. 1987. Underground Storage Tank Corrective Action Technologies. EPA/624/16-87/015.

EPA. 1988. Seminar Proceedings of the Underground Environment of an UST Motor Fuel Release. Office of Research and Development, USEPA, Edison, NJ.

Foglio, J.C., and C.L. Eidam. 1988. Innovative Application of Soil Vapor Extraction Eliminates Off-site Disposal Requirement. Proceedings of the 8th National Conference (Superfund '88). Washington, D.C. Nov. 28-30.

HazTech News. 1989. Soil Biofiltration of VOCs. 4(9):64.

Hinchee, R.E., H.J. Reisinger, D. Burrs, B. Marks, and J. Stepek. 1986. Underground Fuel Contamination, Investigation and Remediation. A Risk Assessment Approach to How Clean is Clean. Proceedings of Petroleum Hydrocarbons and Organic Chemicals in Groundwater: Prevention, Detection and Restoration. Houston, TX.

Hoag, G.E. and B. Cliff. 1985. The Use of the Soil Venting Technique for the Remediation of Petroleum Contaminated Soils. University of Massachusetts Conference. Amherst, MA. October.

Hutzler, N.J., B.E. Murphy, and J.S. Gierke. 1988. State of Technology Review, Soil Vapor Extraction Systems. Hazardous Waste Engineering Research Laboratory, Office of Research and Development, U.S. EPA. Cincinnati, OH.

Jafek, B. 1986. VOC Air Stripping Cuts Costs. Waste Age, October, pp. 66-69.

Johnson, P.C., M.W. Kemblowski, and J.D. Colthart. 1988. Practical Screening Models for Soil Venting Applications. Proceedings of the NWWA/API Conference on Petroleum Hydrocarbons and Organic Chemicals in Groundwater. Houston, TX. pp. 521-546.

Jury, W.A. and R.L. Valentine. 1986. Transport Mechanisms and Loss Pathways for Chemicals in Soil. In: Vadose Zone Modeling of Organic Pollutants. S.C. Hern and S. M. Melacon (eds.) Chelsea, MI: Lewis Publishers. pp. 37-55.

Karimi, A.A., W.J. Farmer, and M.M. Cliath. 1987. Vapor-phase Diffusion of Benzene in Soil. Journal of Environmental Quality. 16(1):38-43.

Kerfoot, H.B. 1988. Is Soil-Gas Analysis an Effective Means of Tracking Contaminant Plumes in Groundwater? What Are The Limitations of the Technology Currently Employed? Ground Water Monitoring Review. 8(2):54-57.

Kerfoot, W.B. 1990. Soil Venting With Pneumatically-Installed Shield Screens. Proceedings of the Fourth National Outdoor Action Conference on Aquifer Restoration, Ground Water Monitoring, and Geophysical Methods. NWWA, Las Vegas, NV.

Knieper, L.H. 1988. VES Cleans Up Gasoline Leak. Pollution Engineering. 20(8):56.

Koltuniak, D.L. 1986. In-situ Air Stripping Cleans Contaminated Soil. Chemical Engineering. 93(16):30-31.

Kreamer, D.K., E.P. Weeks, and G.M. Thompson. 1988. A Field Technique to Measure the Tortuosity and Sorption-Affected Porosity for Gaseous Diffusion of Materials in the Unsaturated Zone with Experimental Results from Near Barnwell, South Carolina. Water Resources Research. 24(4):331-341.

Legiec, I.A. and D.S. Kosson. 1988. In-situ Extraction of Industrial Sludges. Environmental Progress. 7(4):270-278.

Lord, A.E., Jr., D.E. Hullings, R.M. Koerner, and J.E. Brugger. 1990. Vacuum-Assisted Steam Stripping to Remove Pollutants from Contaminated

Soil: A Laboratory Study. Presented at the 16th Annual EPA Hazardous Waste Research Symposium. Cincinnati, OH. April 3-5.

Malmanis, E., D.W. Fuerst, and R.J. Piniewski. 1989. Superfund Site Soil Remediation Using Large-Scale Vacuum Extraction. Proceedings of the 6th National Conference on Hazardous Wastes and Hazardous Materials. New Orleans, LA. April 12-14. pp. 538-541.

Malot, J.J. 1985. Unsaturated Zone Monitoring and Recovery of Underground Contamination. Fifth National Symposium on Aquifer Restoration and Ground Water Monitoring. Columbus, OH. May 21-24.

Malot, J.J., J.C. Agrelot, and M.J. Visser. 1985. Vacuum: Defense System for Ground Water VOC Contamination. Fifth National Symposium on Aquifer Restoration and Ground Water Monitoring. Columbus, OH. May 21-24.

Markley, D.E. 1988. Cost Effective Investigation and Remediation of Volatile-Organic Contaminated Sites. Proceedings of the 5th National Conference on Hazardous Wastes and Hazardous Materials, HMCRI. Las Vegas, NV. April 19-21.

Massmann, J.W. 1989. Applying Groundwater Flow Models in Vapor Extraction System Design. Journal of Environmental Engineering. 115(1):129-149.

Masten, C.J. 1989. Feasibility of Using Ozone in Place of Air in Vapor Stripping Systems. Presented at the Workshop on Soil Vacuum Extraction, R.S. Kerr Environmental Research Laboratory, Ada, OK. April 27-28.

Miller, G.C., V.R. Hebert, and R.G. Zopp. 1987. Chemistry and Photochemistry of Low-Volatility Organic Chemicals on Environmental Surfaces. Environmental Science & Technology. 210(2):1164-1167.

Munz, C. and P.V. Roberts. 1987. Air-Water Phase Equilibria of Volatile Organic Solutes. JAWWA, 79(5):62-69.

Nunno, T.J., J.A. Hyman, and T. Peiffer. 1989. European Approaches to Site Remediation. Hazardous Materials Control. 2(5):38-45.

Oak Ridge National Laboratory. 1987. Draft: Preliminary Test Plan, In-situ Soil Venting Demonstration, Hill AFB, Utah. U.S. Air Force Engineering and Services Center, Tyndall AFB, FL.

Ostendorf, D.W. and D.H. Kampbell. 1989. Biodegradation of Hydrocarbon Vapors in the Unsaturated Zone. Presented at the Workshop on Soil Vacuum Extraction, R.S. Kerr Environmental Research Laboratory, Ada, OK. April 27-28.

Payne, F.C., C.P. Cubbage, G.L. Kilmer, and L.H. Fish. 1986. In-situ Removal of Purgeable Compounds from Vadose Zone Soils. Purdue Industrial Waste Conference. Purdue, IN. May 14.

Radian Corporation. 1989. Short-term Fate and Persistence of Motor Fuels in Soils. Prepared for American Petroleum Institute, Washington, DC.

Rollins, Brown, and Gunnell, Inc. 1985. Subsurface Investigation and Remedial Action, Hill AFB JP-4 Fuel Spill, Provo, Utah. Hill AFB, UT.

Sleep, B.E. and J.F. Sykes. 1989. Modeling the Transport of Volatile Organics in Variably Saturated Media. Presented at the Workshop on Soil Vacuum Extraction. R.S. Kerr Environmental Research Laboratory, Ada, OK, April 27-28.

Smith, C.W., M.L. Kiefer, and R.F. Skach. 1989. Expedited Response Action to Remediate VOC Contamination. Proceedings of the 6th National Conference on Hazardous Wastes and Hazardous Materials. New Orleans, LA. pp. 188-192.

Spencer, W.F., M.M. Cliath, W.A. Jury, and Liam-Zhang. 1988. Volatilization of Organic Chemicals from Soil as Related to Their Henry's Law Constants. Journal of Environmental Quality. 17(3):504-509.

Sterrett, R.J. 1989. Analysis of In-situ Soil Air Stripping Data. Presented at the Workshop on Soil Vacuum Extraction. R.S. Kerr Environmental Research Labortory, Ada, OK. April 27-28.

Stiver, W., W.Y. Shiu, and D. Mackay. 1989. Evaporation Times and Rates of Specific Hydrocarbons in Oil Spills. Environmental Science & Technology. 23(1).

Stover, E.L. 1989. Coproduced Groundwater Treatment and Disposal Options During Hydrocarbon Recovery Operations. Ground Water Monitoring Review. 9(1).

Suenz, G.H., R. Fuentes, and N.E. Pingitore. 1988. A Discriminating Method for the Identification of Soils and Groundwater Contaminated by Hydrocarbons. Proceedings of the Petroleum Hydrocarbons and Organic Chemicals in Ground Water: Prevention, Detection, and Restoration. Houston, TX. November 9-11.

Sykes, J.F. 1989. Modeling the Transport of Volatile Organics in Variably Saturated Media. Presented at the Workshop on Soil Vacuum Extraction, R.S. Kerr Environmental Research Laboratory, Ada, OK. April 27-28.

Terra Vac, Inc. 1987. Union 76 Gas Station Clean-up, Bellview, Florida. Florida Department of Environmental Regulation, Tallahassee, FL.

Texas Research Institute. 1986. Examination of Venting for Removal of Gasoline Vapors From Contaminated Soil. American Petroleum Institute. March, 1980 (Reprinted in 1986).

Thibodeaux, L.J. 1988. Equilibrium and Transport Processes of Vapor Migration in Subterranean Context. In: Seminar Proceedings of the Underground Environment of an UST Motor Fuel Release, Office of Research and

Development, Edison, NJ. pp. 84-90.

Thornton, S.J., and W.L. Wootan. 1982. Venting for the Removal of Hydrocarbon
     Vapors from Gasoline Contaminated Soil. Journal of Environmental
     Science and Health. A17 (1):31-34.

Thornton, S.J., R.E. Montgomery, T. Voynick, and W.L. Wootan. 1984. Removal
     of Gasoline Vapor from Aquifers by Forced Venting. Proceedings of the
     1984 Hazardous Materials Spills Conference, Nashville, TN. pp. 279-286.

Tillman, N., K. Ranlet, and T.J. Meyer. 1989. Soil Gas Surveys: Part 1.
     Pollution Engineering. 21(7):186-189.

Towers, D.S., M.J. Dent, and D.G. Van Arnam. 1988. Evaluation of In-situ
     Technologies for VHO Contaminated Soil. Proceedings of the 5th National
     Conference on Hazardous Wastes and Hazardous Materials, HMCRI, Las
     Vegas, NV. April 19-21.

Treweek, G.P., and J. Wogec. 1988. Soil Remediation by Air/Steam Stripping.
     Proceedings of the 5th National Conference on Hazardous Wastes and
     Hazardous Materials, HMCRI, Las Vegas, NV. April 19-21.

U.S. Patent Office. 1986. Removal of Volatile Contaminants from the   Vadose
     Zone of Contaminated Groundwater. Patent No. 4,593,760.

Walton, J.C., R.G. Baca, J.B. Sisson, and T.R. Wood. 1990. Application of
     Soil Venting at a Large Scale: A Data and Modeling Analysis.
     Proceedings of the Fourth National Outdoor Action Conference on Aquifer
     Restoration, Ground Water Modeling and Geophysical Methods. NWWA, Las
     Vegas, NV.

Wark, K., and C.F. Warner. 1981. Air Pollution: Its Origin and Control. 2nd
     Edition. New York: Harper & Row Publishers.

Water Resources Associates, Inc. 1989. Company literature. Austin, TX.

Wenck Associates, Inc. 1985. Project Documentation: Work Plan, ISV/In-situ
     Volatilization, Sites D and G, Twin Cities Army Ammunition Plan, Federal
     Cartridge Corporation, New Brighton, MN.

Weston, Roy F., Inc. 1985. Appendices - Task 11, In-situ Solvent Stripping
     from Soils Pilot Study. U.S. Army Toxic and Hazardous Materials Agency,
     Aberdeen Providing Grounds, MD.

Wilson, D.E., R.E. Montgomery, and M.R. Sheller. 1987. A Mathematical Model
     for Removing Volatile Subsurface Hydrocarbons by Miscible Displacement.
     Water, Air, and Soil Pollution. 33(3-4):231-235.

Wilson, D.J., R.D. Mutch, Jr. and A.N. Clarke. 1989. Modeling of Soil Vapor
     Stripping. Presented at the Workshop on Soil Vacuum Extraction, R.S.
     Kerr Environmental Research Laboratory, Ada, OK. April 27-28.

Woodward—Clyde Consultants. 1985.   Performance Evaluation Pilot Scale
    Installation and Operation Soil Gas Vapor Extraction System. Time
    Oil Company Site. Tacoma, Washington.

Yaniga, P.M., and W. Smith. 1985.   Aquifer Restoration: In—situ
    Treatment and Removal of Organic and Inorganic Compounds.
    Groundwater Contamination and Reclamation. August. pp. 149—165.

Appendix A
Review of Soil Vapor Extraction System Technology

Neil J. Hutzler[a], Blaine E. Murphy[b], John S. Gierke[a]

ABSTRACT

Soil vapor extraction is a cost-effective technique for the removal of volatile organic chemicals (VOCs) from contaminated soils. Soil air extraction processes cause minimal disturbance of the contaminated soil and can be constructed from standard equipment. There is demonstrated experience with soil vapor extraction at pilot- and field-scale, the process can be used to treat larger volumes of soil than can be practically excavated, and there is a potential for product recovery.

A soil vapor extraction system involves extraction of air containing volatile chemicals from unsaturated soil. Fresh air is injected or flows into the subsurface at locations around a spill site, and the vapor-laden air is withdrawn under vacuum from recovery or extraction vents. A typical system consists of: (1) one or more extraction vents, (2) one or more air inlet or injection vents (optional), (3) piping or air headers, (4) vacuum pumps or air blowers, (5) flow meters and controllers, (6) vacuum gauges, (7) sampling ports, (8) air/water separator (optional), (9) vapor treatment (optional), and (10) a cap (optional).

A large number of pilot- and full-scale soil vapor extraction systems have been constructed and studied under a wide range of conditions. Based on a review of 17 studies, a number of conclusions can be drawn. Soil vapor extraction can be effectively used to remove a wide range of volatile chemicals over a wide range of conditions. The design and operation of these systems is flexible enough to allow for rapid changes in operation, thus, optimizing the removal of chemicals. Air injection and the capping of a site can control air movement, but injection systems need to be carefully designed to avoid spreading contamination. Incremental installation of vents, while probably more expensive, allows for a greater degree of freedom in design. While a number of variables intuitively affect the rate of chemical extraction, no extensive study to correlate variables to extraction rates has been identified.

---

[a]Department of Civil and Environmental Engineering, Michigan Technological University, Houghton, Michigan 49931
[b]Woodward-Clyde Consultants, Chicago, IL 60603

INTRODUCTION

Soil may become contaminated with volatile organic chemicals such as industrial solvents and gasoline components in a number of ways. The sources of contamination at or near the earth's surface include intentional disposal, leaking underground storage tanks, and accidental spills. Contamination of groundwater from these sources can continue even after discharge has stopped, because the unsaturated zone above a groundwater aquifer can retain a portion or all of the contaminant discharge. As rain infiltrates, chemicals are washed from the contaminated soil and migrate towards groundwater.

Alternatives for decontaminating unsaturated soil include excavation with on-site or off-site treatment or disposal, biological degradation, and soil washing. Soil vapor extraction is also an accepted, cost-effective technique for the removal of volatile organic chemicals (VOCs) from contaminated soils (Bennedsen, 1987; Malot and Wood, 1985; Payne et al., 1986). With vapor extraction, it is possible to clean up spills before the chemicals reach the groundwater table. Soil vapor extraction technology is often used in conjunction with other clean-up technologies to provide complete restoration of contaminated sites (Malot and Wood, 1985; Oster and Wenck, 1988; CH_2M-Hill, 1987).

Unfortunately, there are few guidelines for the optimal design, installation, and operation of soil vapor extraction systems (Bennedsen, 1987). Theoretically-based design equations which define the limits of this technology are especially lacking. Because of this, the design of these systems is mostly empirical. Alternative designs can only be compared by the actual construction, operation, and monitoring of each design.

One of the major objectives of this paper is to critically review information that describes current practices. The information is summarized in several tables, which form the basis for a discussion of the design, installation, and operation of these systems.

SCOPE OF STUDY

As a part of this investigation, information on 7 pilot-scale, and 10 full-scale studies have been reviewed with respect to the design and operational variables. These sites along with their location, the study type, the duration of study or date the study began, and the project status are listed in Table A-1. While this list is by no means complete, it provides a means for discussing the range of system variables for current installations.

This technology has been referred to by several names, including "subsurface venting", "vacuum extraction", "in situ soil air stripping", and "soil venting", as well as "soil vapor extraction". The term "soil vapor extraction" is used throughout this paper. Soil vapor extraction technology seems rather simple in concept, but its application appears to be relatively recent as indicated by the dates of the available reports. There is a wide variety of system designs and operating conditions.

Table A-1.   List of Typical Pilot and Field Soil
Vapor Extraction Systems.

SITE	LOCATION	STUDY SCALE	DATE OR DURATION	1988 STATUS	REFERENCES	NAME USED FOR SYSTEM
FUEL MARKETING TERMINAL	Granger Indiana	pilot	12 days 10 days 15 days	completed	Crow et al., 1987 Amer Petr Inst, 1985	Subsurface Venting
VALLEY MANUFACTURING	Groveland Massachusetts	pilot	Jan-Apr 88	completed	Enviresponse, 1987	Vacuum Extraction
INDUSTRIAL TANK FARM	San Juan Puerto Rico	pilot/ full	30 months	completed?	Malot & Wood, 1985 Malot, 1985	Vacuum Extraction
TIME OIL COMPANY	Tacoma Washington	pilot/ full	11 days (Aug 1985)	pilot completed	Woodward-Clyde, 1985	Soil Gas Vapor Extraction
SOLVENTS STORAGE TANK	Cupertino California	pilot/ full	several months	completed?	Bennedsen, 1987	Soil Gas Vapor Extraction
TCAAP PILOT 1	New Brighton Minnesota	pilot	67 days	completed	Anastos et al., 1985	In-situ Venting
TCAAP PILOT 2	"	pilot	78 days	completed	Anastos et al., 1985	"
TCAAP SITE D	"	full	Feb 1986	ongoing	Wenck, 1985 Oster & Wenck, 1988	"
TCAAP SITE G	"	full	Feb 1986	ongoing	Wenck, 1985 Oster & Wenck, 1988	"
GAS STATION	unknown	full	?	completed?	Malot & Wood, 1985	Vacuum Extraction
UNION 76 GAS STATION	Bellview Florida	full	7 months	ongoing	Camp, Dresser, & McKee, 1987, 1988	Vacuum Extraction
SOUTH PACIFIC RAILROAD	Benson Arizona	full	7 months	completed	Johnson, 1988 Johnson & Sterrett, 1988	In-situ Soil Air Stripping
CUSTOM PRODUCTS	Stevensville Michigan	full	Dec 1988 >280 days	completed?	Payne et al., 1986 Payne & Lisiecki, 1988	Forced Air Circulation
ELECTRONIC MANUFACTURING	Santa Clara Valley, CA	full	3 yrs	completed?	Bennedsen, 1985	Vapor Extraction
PAINT STORAGE	Dayton Ohio	full	since July 1987	ongoing	Payne & Lisiecki, 1988	Enhanced Volatilization
THOMAS SOLVENT COMPANY	Battle Creek Michigan	full	since Jan 1988	ongoing	CH_2M-Hill, 1987	Vacuum Extraction
HILL AFB VERTICAL VENTS	Hill AFB	full	since Fall 1988	ongoing	Oak Ridge National Lab, 1988 Radian, Corp., 1987	Soil Venting
HILL AFB LATERAL SYSTEM	(3 parallel) (extraction) (systems)	"	"			"
HILL AFB SOIL PILE		"	"			"

From Table A-1, it can be seen that soil vapor extraction systems have
been installed at locations across the United States and have been observed
over periods ranging from several weeks to several years. Projects ranging in
status from being complete to being in the preliminary design stage have been
identified. Some of the studies were too short to fully assess the
effectiveness of this technology.

PROCESS DESCRIPTION

A soil vapor extraction system involves the extraction of air containing
volatile chemicals from unsaturated soil. Fresh air is injected or flows into
the subsurface at locations around a spill site, and the vapor-laden air is
withdrawn under vacuum from recovery or extraction vents. A typical soil
vapor extraction system, such as the one shown in Figure A-1, consists of: (1)
one or more extraction vents, (2) one or more air inlet or injection vents
(optional), (3) piping or air headers, (4) vacuum pumps or air blowers, (5)
flow meters and controllers, (6) vacuum gauges, (7) sampling ports, (8)
air/water separator (optional), (9) vapor treatment (optional), and (10) a cap
(optional). The design of each of these components is discussed below.

SYSTEM VARIABLES

A number of variables characterize the successful design and operation of
a vapor extraction system. They may be classified as site conditions, soil
properties, chemical characteristics, control variables, and response
variables (Anastos et al., 1985; Enviresponse, 1987). Table A-2 lists
specific variables that belong to these groups.

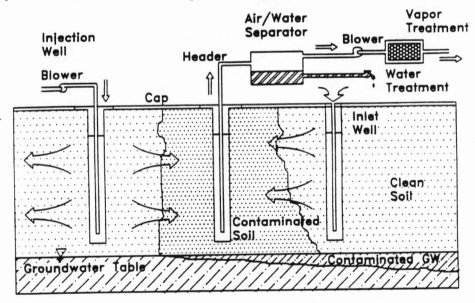

Figure A-1. Soil Vapor Extraction System.

Site Conditions

Most site conditions can not be changed. The extent to which VOCs are dispersed in the soil, vertically and horizontally, is an important consideration in deciding if vapor extraction is preferable to other methods. Soil excavation and treatment is probably more cost effective when only a few hundred cubic yards of near-surface soils are contaminated (Bennedsen, 1987). If the spill has penetrated more than 20 or 30 feet or has spread through an area over several hundred square feet at a particular depth or if the spill volume is in excess of 500 cubic yards, then excavation costs begin to exceed those associated with a vapor extraction system (CH$_2$M Hill, 1985; Payne et al., 1986).

TABLE A-2.   SOIL VAPOR EXTRACTION SYSTEM VARIABLES.

Site Conditions	Control Variables
Distribution of VOCs	Air withdrawal rate
Depth to groundwater	Vent configuration
Infiltration rate	Extraction vent spacing
Location of heterogeneities	Vent spacing
Temperature, humidity	Ground surface covering
Atmospheric pressure	Pumping duration
Location of structures	Inlet air VOC concentration
Rainfall	and moisture content
Barometric pressure	
Soil Properties	**Response Variables**
Permeability (air and water)	Pressure gradients
Porosity	Final distribution of VOCs
Organic carbon content	Final moisture content
Soil structure	Extracted air concentration
Soil moisture characteristics	Extracted air moisture
Particle size distribution	Extracted air temperature
	Power usage
Chemical Properties	
Henry's constant	
Solubility	
Adsorption equilibrium	
Diffusivity (air and water)	
Density	
Viscosity	

The depth to groundwater is also important. Where groundwater is at depths of more than 40 feet and the contamination extends to the groundwater, use of soil vapor extraction systems is one of the few ways to remove VOCs from the soil (Malot and Wood, 1985). Groundwater depth in some cases may be lowered to increase the volume of the unsaturated zone.

Heterogeneities influence air movement as well as the location of chemical, making it more difficult to position extraction and inlet vents. There generally will be significant differences in the air conductivity of the various strata of a stratified soil. A horizontally-stratified soil may be favorable for vapor extraction because the relatively impervious strata will limit the rate of vertical inflow from the ground surface and will tend to extend the influence of the applied vacuum horizontally from the point of extraction.

The specific location of the contaminant on a property and the type and extent of development in the vicinity of the contamination may favor the installation of a soil vapor extraction system. For example, if the contamination extends across property lines, beneath a building or beneath an extensive utility trench network, vapor extraction should be considered.

Temperature affects the performance of soil vapor extraction systems primarily because of its influence on chemical properties such as Henry's constant, solubility, and sorption capacity. In most cases, extraction systems are operated at ambient temperatures.

Soil Properties

The soil characteristics at a particular site have a significant effect on the applicability of vapor extraction systems. Air conductivity controls the rate at which air can be drawn from soil by the applied vacuum. Grain size, moisture content, soil aggregation, and stratification are probably the most important properties (Bennedsen et al., 1985; Hutzler et al., 1988). The soil moisture content or degree of saturation is also important in that it is easier to draw air through drier soils. As the size of a soil aggregate increases, the time required for diffusion of the chemical out of the immobile regions also increases. However, even clayey or silty soils may be effectively ventilated by the usual levels of vacuum developed in a soil vapor extraction system (Camp, Dresser, and McKee, 1987; Terra Vac, 1986b). The success of the soil vapor extraction in these soils may depend on the presence of more conductive strata, as would be expected in alluvial settings, or on relatively low moisture contents in the finer-grained soils.

Chemical Properties

In conjunction with site conditions and soil properties, chemical properties will dictate whether a soil vapor extraction system is feasible. A vapor-phase vacuum extraction system is most effective at removing compounds that exhibit significant volatility at the ambient temperatures in soil, compounds exhibiting vapor pressures over 0.5 mm of mercury (Bennedsen et al., 1985) and compounds which have values of dimensionless Henry's Law constants greater than 0.01. Compounds which have been effectively removed by vapor extraction include trichloroethene, trichloroethane, tetrachloroethene, and most gasoline constituents. Compounds which are less easily removed include trichlorobenzene, acetone, and heavier petroleum fuels (Payne et al., 1986; Bennedsen et al., 1985; Texas Research Institute, 1980).

Soluble compounds tend to travel farther in soils where the infiltration rate is high. The movement of chemicals with affinity for soil organic material or mineral adsorption sites will be retarded. In drier soils, chemical density and viscosity have the greatest impact on organic liquid movement, however, in most current systems, the contamination is old enough that no further movement of free product occurs.

Control Variables

Soil vapor extraction processes are flexible in that several variables can be adjusted during design or operation. Higher air flow rates tend to increase vapor removal because the zone of influence is increased and air is forced through more of the air-filled pores. More vents allow better control of air flow but also increase construction and operation costs. The water infiltration rate can be controlled by placing an impermeable cap over the site. Intermittent operation of the blowers allows time for chemicals to diffuse from immobile water and air and permits removal at higher concentrations.

Response Variables

Parameters responding to soil vapor extraction system performance include: air pressure gradients, VOC concentrations, moisture content, and power usage. The rate of vapor removal is expected to be primarily affected by the chemical's volatility, its sorptive capacity onto soil, the air flow rate, the distribution of air flow, the initial distribution of chemical, soil stratification or aggregation, and the soil moisture content.

SOIL VAPOR EXTRACTION SYSTEM DESIGN

Tables A-3, A-4, and A-5 summarize the design and operation of the major components of the pilot- and field-scale systems reviewed for this paper. These include extraction vent design and placement, piping and blower systems and the miscellaneous components discussed previously.

Vent Design and Placement

Extraction vents --

Typically, extraction vents are designed to fully penetrate the unsaturated soil zone or the geologic stratum to be cleaned. An extraction vent usually is constructed of slotted plastic pipe placed in a permeable packing as shown in Figure A-2. Vents may be installed vertically or horizontally. Vertical alignment is typical for deeper contamination zones and results in radial flow patterns. If the depth of the contaminated soil or the depth to the groundwater table is less than 10 to 15 feet, it may be more practical to dig a trench across the area of contamination and install

Table A-3. Pilot and Field Soil Vapor Extraction Systems --
Vent Design and Placement.

SITE	EXTRACTION VENTS				AIR INPUT		
	NUMBER AND TYPE	VENT MATERIAL	VENT CONSTRUCTION	VENT SPACING	NUMBER AND TYPE	VENT MATERIAL	VENT CONSTRUCTION
FUEL MARKETING TERMINAL	2 vents	2" PVC	screened 14 to 20 ft BLS	20, 40, & 100 ft	4 air inlet vents	2" PVC	screened 14 - 20 ft BLS
VALLEY MANUFACTURING	8 vents 4 sh, 4 deep	4" PVC	up to 30 ft deep	20 ft	surface	na	na
INDUSTRIAL TANK FARM	3 vents	?	25 to 75 ft BLS & at 300 ft BLS	?	surface	na	na
TIME OIL COMPANY	7 vents	2" PVC	screened 6 to 25 ft BLS	40-90 ft	surface	na	na
SOLVENTS STORAGE TANK	1 vent	?	?	na	1 air inlet vent	?	?
TCAAP PILOT 1	9 vent grid	3" PVC	grav. pack, slotted 5 to 20 ft BLS	20 ft	4 vents	3" PVC	slotted 15 - 20 ft BLS
TCAAP PILOT 2	9 vent grid	3" PVC	grav. pack, slotted 5 to 20 ft BLS	50 ft	4 vents	3" PVC	slotted 15 - 20 ft BLS
TCAAP SITE D	39 vents	3" PVC	grav. pack, slotted 5 to 25 - 35 ft BLS	25 ft	surface or air inlet	vents can be air inlets	same as extraction
TCAAP SITE G	89 vents	3" PVC	grav. pack, slotted 5 - 25 to 35 ft BLS	25 ft	surface or air inlet	vents can be air inlets	same as extraction
GAS STATION	vertical & horizontal	?	?	?	surface	na	na
UNION 76 GAS STATION	6 vents 3 sh, 3 deep	4" PVC	slotted 10 to 15 ft	14-50 ft	surface	na	na
SOUTH PACIFIC RAILROAD	79 vents	2" PVC	15 to 25 ft deep	variable	surface & injection	21 vents were used as AIV	same as extraction
CUSTOM PRODUCTS	1 vent	2" galv. steel	gravel pack 8 to 25 ft BLS	50-70 ft	6 air inj. vents	1.25" PVC	gravel pack 15 to 25 ft BLS
ELECTRONIC MANUFACTURING	1 to 2 vents	2" diam.	?	?	1 to 2 air inlets	2" diam.	?
PAINT STORAGE	over 20 vents	galv. steel	?	?	large no. of vents	poly-ethylene	?
THOMAS SOLVENT COMPANY	14 vents	4" PVC	?	?	surface	na	na
HILL AFB VERTICAL VENTS	15 vertical vents	4" PVC	screened 10 to 30 ft BLS	20 and 40 ft	surface or air inlet	vents can be air inlets or injection	same as extraction
HILL AFB LATERAL SYSTEM	6 laterals	4" poly-ethylene	20 ft BLS	15 ft	surface or air inlet	laterals can be inlets	same as extraction
HILL AFB SOIL PILE	8 laterals	4" poly-ethylene	5 ft above pile bottom	18 ft	surface	na	na

AIV -- air inlet vent   GWT -- ground water table   ? -- no information
BLS -- below land surface   na -- not applicable   sh -- shallow

Table A-4.   Pilot and Field Soil Vapor Extraction Systems --
Piping and Blower Systems.

SITE	PIPING	VACUUM SOURCE	AIR FLOW	VACUUM	GAS FLOW METER
FUEL MARKETING TERMINAL	1 & 2" PVC	2 liquid ring vacuum pumps	23 cfm 18 cfm 40 cfm	0.4" Hg 0.3" Hg 0.9" Hg	pitot tube w/ diff. press. meas
VALLEY MANUFACTURING	PVC manifold heated	blower	3 to 800 cfm?	0-29" Hg	X
INDUSTRIAL TANK FARM	?	vacuum pump	18 cfm 150 cfm	25-30" Hg	?
TIME OIL COMPANY	2" PVC manifold	blower	210 cfm 30 cfm/well	?	pitot tube w/ diff. press. meas
SOLVENTS STORAGE TANK	?	blowers	10 cfm 100 cfm	0.24" Hg 6" Hg	?
TCAAP PILOT 1	3" PVC grid insulated	2 blowers 1 extr., 1 inj.	40 - 55 cfm	?	X
TCAAP PILOT 2	3" PVC grid insulated	2 blowers 1 extr., 1 inj.	200-220 to 100 to 50 cfm	?	X
TCAAP SITE D	8 to 18" steel insul. manifold heated	up to 4 blowers variable speed	2200 cfm per blower	1.8" Hg	totalizing flow meter
TCAAP SITE G	12 to 24" steel insul. manifold heated	up to 4 blowers variable speed	5700 cfm per blower	1.8" Hg	totalizing flow meter
GAS STATION	?	vacuum pump	?	?	?
UNION 76 GAS STATION	manifold	vacuum pump	?	?	?
SOUTH PACIFIC RAILROAD	4" PVC manifold	3 blowers separate systems	86 - 250 cfm	0.7-0.6" Hg	none
CUSTOM PRODUCTS	2" galv. steel	rotary vane vac. pump	10.2 cfm	4.5" Hg	X
ELECTRONIC MANUFACTURING	duct	2 blowers	10 cfm to 100 cfm	0.2 to 3" Hg	?
PAINT STORAGE	galv. st., heat manifolds	8 blowers	?	?	?
THOMAS SOLVENT COMPANY	?	blower	?	?	?
HILL AFB VERTICAL VENTS	10-16" metal manifold	common source	up to 2000 cfm	up to 8" Hg	orifices with Magnehelic differential pressure gauges or U-tube manometers
HILL AFB LATERAL SYSTEM	same	2 rotary lobe blowers up to 1000 cfm each		1 to 2"Hg normal	
HILL AFB SOIL PILE	same	250 cfm aux. blower			

X -- listed component present, no detailed information
? -- no information

Table A-5. Pilot and Field Soil Vapor Extraction Systems --
Miscellaneous Components.

SITE	IMPERMEABLE CAP	AIR/WATER SEPARATOR	VAPOR TREATMENT	GAUGES	SAMPLING PORTS	TYPES OF MONITORING
FUEL MARKETING TERMINAL	plastic membrane	none	none	vacuum temperature	vent heads exhaust port	monitoring well vapor probes
VALLEY MANUFACTURING	none	500 gallon	GAC	vacuum	vent head system lines	exhaust gas monitoring wells
INDUSTRIAL TANK FARM	none	condenser	recovery tank	?	?	exhaust gas monitoring wells
TIME OIL COMPANY	none	55 gallon tank	none	vacuum temperature	vent heads exhaust port	soil borings exhaust gas
SOLVENTS STORAGE TANK	none	55 gallon tank	none	?	?	exhaust gas
TCAAP PILOT 1	?	none	GAC	vacuum temperature	inlet ports exhaust port	soil borings air monitoring
TCAAP PILOT 2	?	none	GAC	vacuum temperature	inlet ports exhaust port	soil borings air monitoring
TCAAP SITE D	18" clay	none	none	vacuum	vent heads central header	soil vapor air monitoring exhaust gas
TCAAP SITE G	18" clay	none	none/GAC	vacuum	vent heads central header	soil vapor air monitoring exhaust gas
GAS STATION	concrete pavement	condenser?	none	?	?	monitoring wells
UNION 76 GAS STATION	existing pavement	gas/water separator	none	?	?	monitoring wells soil borings vapor probes
SOUTH PACIFIC RAILROAD	none	none	none	?	?	monitoring wells soil borings
CUSTOM PRODUCTS	6 mil-poly-ethylene	liquid trap	GAC	vacuum	before and after GAC	exhaust gas soil samples
ELECTRONIC MANUFACTURING	none	none	none	vacuum	exhaust	exhaust gas
PAINT STORAGE	clay cover & concrete	trap w/pump to tank	combustion	vacuum temperature	vent heads vapor, water	monitoring wells soil borings
THOMAS SOLVENT COMPANY	none	none	GAC	vacuum temperature	exhaust GAC outlets	monitoring wells soil borings air monitoring
HILL AFB VERTICAL VENTS	80'x 140' plastic	160 gallon knock-out drum	2 catalytic incinerators 1 - 500cfm	vacuum temperature humidity	vent heads exhaust	pressure monitoring wells
HILL AFB LATERAL SYSTEM	concrete tank pad	"	fluidized bed 1 - 1000cfm fixed bed	"	"	soil borings
HILL AFB SOIL PILE	none	"		"	"	

GAC -- granular activated carbon
? -- no information

perforated piping in the trench bottom versus installing vertical extraction
vents (Oak Ridge National Lab, 1988; Connor, 1988). The surface of the
vertical vents or the trench for horizontal vents is usually grouted to
prevent the direct inflow of air from the surface along the vent casing or
through the trench. Usually several vents are installed at a site, especially
if soil strata are highly variable in terms of permeability. In stratified
systems, more than one vent may be installed in the same location, each
venting a given strata (Camp, Dresser, and McKee, 1987, 1988). Extraction
vents can be installed incrementally starting with installation in the area of
highest contamination (Payne and Lisiecki, 1988; Johnson and Sterrett, 1988).
This allows the system to be brought on-line as soon as possible.

Vent spacing is usually based on an estimate of the radius of influence
of an individual extraction vent (Malot and Wood, 1985; Wenck, 1985; Oak Ridge
National Lab, 1988). In the studies reviewed, vent spacing has ranged from 15
to 100 feet. Johnson and Sterrett (1988) suggest that vent spacing should be
decreased as soil bulk density increases or the porosity of the soil
decreases.

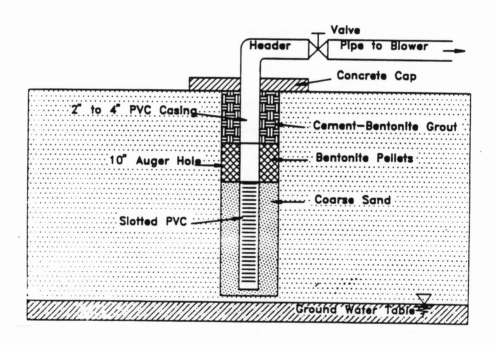

Figure A-2. Typical Extraction/Air Inlet Vent Construction.

One of the major differences noted between systems was the soil boring diameter. Larger borings are preferred to provide air/water separation within the packing.

Air Input

In the simplest soil vapor extraction systems, air flows to an extraction vent from the ground surface as depicted in Figure A-3. To enhance air flow through zones of maximum contamination, it may be desirable to include air inlet vents in the installation. Injection or inlet vents may be located at numerous places around the site. The function of inlet vents and caps is to control the flow of air into a contaminated zone. Air vents are passive. Injection vents force air into the ground and can be used in closed-loop systems (Payne et al., 1986). Injection vents are installed at the edge of a site, as depicted in Figure A-4, so as not to force contamination away from the extraction vents. In addition, injection vents are often installed between adjacent extraction points to ensure pressure gradients in the direction of the extraction vents (Payne et al., 1986). Typically, injection and inlet vents are similar in construction to extraction vents. In some installations, extraction vents have been designed so they can also be used as air inlets (Wenck, 1985; Oak Ridge National Lab, 1988).

Usually, only a fraction of extracted air comes from air inlets (American Petroleum Institute, 1985; Crow et al., 1987; Ellgas and Marachi, 1988). This indicates that air drawn from the surface is the predominant source of clean air.

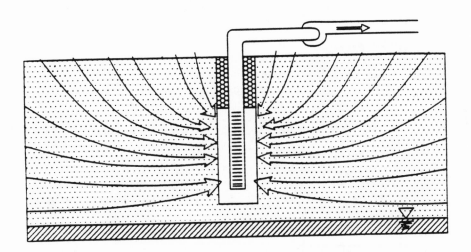

Figure A-3. Air Flow Patterns in Vicinity of a Single
Extraction Vent -- No Cap.

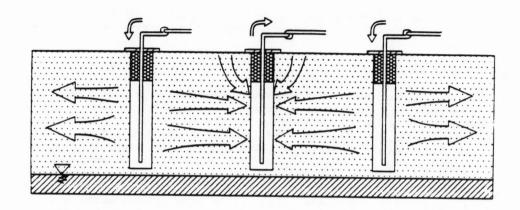

Figure A-4.   Air Flow from Injection Vents.

Thornton et al. (1984) investigated the effects of air flow rate, and the
configuration of the inlet and extraction vents on gasoline recovery from an
artificial aquifer.   They determined that screening geometry only had an
effect at the low air flow rates.   At low flow rates, higher recovery rates
resulted when the screen was placed near the water table versus being screened
the full depth of the aquifer.   A similar assessment was made by Woodward-
Clyde Consultants (1985) at the Time Oil Company site.   Woodward-Clyde
engineers suggested that the vents should be constructed with approximately 20
feet of solid pipe between the top of the screen and the soil surface to
prevent the short circuiting of air and to aid in the extraction of deep
contamination.

Piping and Blower Systems

Table A-4 summarizes information on the design of piping systems and the
selection of blowers for vapor extraction systems.

Piping --
Piping materials connecting the vents to headers as well as the headers
themselves are usually plastic or steel.   Wenck (1985) suggests that headers
be constructed of steel for durability, especially in colder climates.
Headers may be configured as a manifold or in a grid, although, manifold
construction appears to be the most common.   Pipes and headers are usually
buried or wrapped with heat tape and insulated in northern climates to prevent
freezing of condensate (Wenck, 1985).

Valving --
A control/shut-off valve is usually installed at each venthead and at
other critical locations, such as lateral/header connections, to provide
operational flexibility and optimize extraction rates.   Typically, ball or
butterfly valves are used because they provide better flow control.

Vacuum Source --
    The vacuum for extracting soil air is developed by an ordinary positive
displacement industrial blower, a rotary blower, vacuum or aspirator pump, or
a turbine. There are a large number of commercially available blower models.
In the studies reported herein, the blowers have had ratings ranging from 100
to 6,000 cubic feet per minute at vacuums up to about 30 inches Hg gauge as
shown in Table A-4. Ratings of the electric drive motors are usually 10
horsepower or less. The pressure from the outlet side of the pumps or blowers
is usually used to push the exit gas through a treatment system and can be
used to force air back into the ground if injection vents are used (Payne et
al., 1986), although, it is more common to use a separate blower for injection
(Anastos et al., 1985). Vapor treatment efficiency can be improved by
installing the blower between the moisture separator and the vapor treatment
system to take advantage of the heat generated by the blower. The blower or
blowers are usually housed in a temporary building on-site.

Gas Flow Meter --
    A flow meter should be installed to monitor the volume of extracted air.
This measurement is used in conjunction with gas analysis to determine the
total mass of vapor extracted from the soil. Flow measurements from
individual vents are useful for optimizing extraction system operation. A
flowmeter consisting of an orifice plate and manometer, together with the
appropriate rating curve, will yield the system discharge air flow rate.

Miscellaneous Components

    In addition to the basic vent, piping, and blower components, a soil
vapor extraction system may require a cover, air/water separator, and vapor
treatment. Table A-5 summarizes the range of design of miscellaneous
components of the various pilot- and field systems.

Impermeable Cap --
    The surface of the entire site may be sealed with plastic sheeting, clay,
concrete, or asphalt as indicated in Table A-5. If movement of the air toward
the extraction vent is desired to be more radial than vertical, then an
impermeable cap should be added. The cap controls the air flow pathway so
that clean air is more likely to come from air vents or injection vents.
Without the cap (Figure A-3), a more vertical movement of air from the soil
surface takes place. But when an impermeable cap is in place, the radius of
influence around the extraction vent is extended (Figure A-5). Thus, more of
the contaminated soil may be cleansed by the air flow.

    The use of a polyethylene cover will also prevent or minimize
infiltration, which, in turn, reduces the moisture content and further
chemical migration. With little or no infiltration, water is less likely to
be extracted from the system, thus reducing the need for an air/water
separator. In very dry climates, a reduction of moisture content below which
partial drying of the soil occurs, extraction system efficiency may be reduced
due to increased adsorption capacity of the dry soil (Johnson and Sterrett,
1988).

Air/Water Separator --

If water is pulled from the extraction vents, an air/water separator is required to protect the blowers or pumps and to increase the efficiency of vapor treatment systems. The condensate may then have to be treated as a hazardous waste depending on the types and concentrations of contaminants. The need for a separator may be eliminated by covering the treatment area with an impermeable cap. In some cases, a gasoline/water separator may be used in conjunction with a combination vapor extraction/pumping system for gasoline product recovery (Malot and Wood, 1985; Thornton et al., 1984).

Vapor Treatment --

Air emission problems should not be created while solving a soil contamination problem. Vapor treatment may not be required for systems that produce a very low emission rate of easily degradable chemicals. The decision to treat vapor must be made in conjunction with air quality regulators. There are several treatment systems available that limit or control air emissions. These include liquid/vapor condensers, incinerators, catalytic converters, and gas-phase granular activated carbon (GAC).

If air emissions control or vapor treatment is required for an installation, a vapor phase activated carbon adsorber system probably will be the most practical system depending on chemical emission rates and VOC levels, although catalytic oxidation units have produced favorable results (Bennedsen, 1985). Gas-phase GAC may require heating of the extracted air to control the relative humidity in order to optimize the carbon usage rate. As the fraction of water increases, the capacity for the target chemical decreases and the carbon replacement rate increases. The spent carbon may be considered as a hazardous waste depending on the contaminants (Enviresponse, 1987).

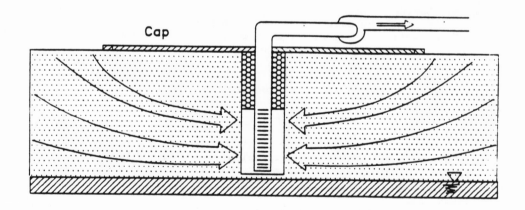

Figure A-5.   Air Flow Patterns with Impermeable Cap in Place.

On one project, where the initial extraction rate of volatiles was over 200 pounds per day, the extracted gas was able to be piped to the combustion air intake zone of a nearby industrial boiler that was in continuous operation (Bennedsen et al., 1985). Laboratory analyses did not detect unwanted volatiles in the boiler emissions. Incineration can be self-sustained combustion if the vapor contains high concentrations of hydrocarbons or combustible volatile chemicals. Usually there is a lag time to achieve a high concentration of combustibles. Concentration of volatiles in the air stream might be increased by intermittent blower operation or by intermittently operating different extraction vents. Some systems have auxiliary fuels to maintain a desired exhaust temperature.

Pitot Tubes and Pressure Gauges --
    Various monitoring devices such as sampling ports, vacuum gauges, and pitot tubes are required for estimating vapor discharges. Pressure gauges are used to monitor the pressure losses in the overall system and to optimize air flows.

Sampling Ports --
    Sampling ports are usually installed at each vent head, at the blower, and after gas treatment. The basic measurements required to assess soil vapor extraction system performance are the system air flow rate and the concentration of volatile organic chemicals in the extracted flow. A gas chromatograph equipped with an appropriate detector for the compounds expected to be present in the exhaust gas is typically used to provide VOC concentration data.

Monitoring Systems --
    Vapor and pressure monitoring probes may be placed in the soil surrounding the extraction system to measure vapor concentrations and the radius of influence of the extraction vents. The monitoring wells are usually required to assess the final clean-up of a particular site.

SITE CONDITIONS

Soil and Geological Conditions

    Table A-6 briefly summarizes the geologic conditions at the various pilot and field sites. Although it has been suggested that soil vapor extraction systems should be used primarily in highly permeable soils, they have been installed in soils with a wide range of permeabilities. The range of areas and volumes of soil vented is large. Soil vapor extraction systems have been used in shallow as well as deep unsaturated zones. Much of the information needed to fully assess the effects of soil properties (moisture content, organic carbon content, and porosity) on vapor extraction is not available.

    As the permeability of the soil decreases, the time required for extraction and decontamination increases. In addition to permeability, the presence of heterogeneities make it more difficult to position inlet and extraction vents. For example, the effect of a clay lens at the Groveland site resulted in perched water table. During high rainfall periods, the contaminant seeped over the lip of this clay lens and spread further.

Table A-6. Pilot and Field Soil Vapor Extraction Systems --
Soil and Geological Conditions.

SITE	SOIL/ GEOLOGY	GWT DEPTH	SOIL POROSITY	HYDRAULIC CONDUCTIVITY	MOISTURE CONTENT	AREA AFFECTED
FUEL MARKETING TERMINAL	sand and fine sand layers w/ traces of clay and silt	25 ft	0.38	10^{-3} cm/s	?	two 60 ft^2 areas
VALLEY MANUFACTURING	5 - 12 ft of sand over 5 - 10 ft of clay over glacial till	27-52 ft	?	permeable to impermeable	perched water	?
INDUSTRIAL TANK FARM	40 - 210 ft clayey silts 900 ft limestone	300 ft	?	very permeable	?	4,400,000 cu yds
TIME OIL COMPANY	sand and gravel with some silt	>30 ft	?	$3x10^{-3}$ cm/s	?	30,000 sq
SOLVENTS STORAGE TANK	unknown	85 ft	?	?	?	unknown
TCAAP PILOT 1	4 - 6 ft sand and loamy sand fill over stained low permeability sediments over sand	170 ft	?	very permeable	?	3800 to 33000 cu yds
TCAAP PILOT 2	same as Pilot 1	"	?	"	?	"
TCAAP SITE D	same as Pilot 1	"	?	"	?	"
TCAAP SITE G	up to 135 ft sand over glacial till and sand	130 ft	?	very permeable	?	?
GAS STATION	6 - 12 ft clayey soil grading to silt & sand	8-10 ft	?	impermeable?	?	?
UNION 76 GAS STATION	18 - 21 ft clayey sand over 5-13 ft gumbo clay over 28-42 ft silty sand over limestone	48-53 ft	?	?	?	unknown
SOUTH PACIFIC RAILROAD	20 - 25 ft silt & sand, gravel layers, 50 ft silty clay @ 40 ft	240 ft	0.1 - 0.3	10^{-4} cm/s	2 - 5%	60 x 70 ft
CUSTOM PRODUCTS	30 ft of fine sand	30 ft	?	?	?	50 acres
ELECTRONIC MANUFACTURING	alluvial clayey silts and sands	90 ft	?	relatively impervious	?	?
PAINT STORAGE	sandy soil with clay strata sands and gravels	40-50 ft	?	"?	?	?
THOMAS SOLVENT COMPANY	a sand and gravel alluvial deposit over sandstone	22 ft	?	sand - 0.1 cm/s bedrock- 0.06 cm/s	?	?
HILL AFB VERTICAL VENTS	4 ft silty sand underlain by 16 - 31 ft of sand underlain by	600 ft	?	permeable to impermeable	perched water	200 x 120 ft
HILL AFB LATERAL SYSTEM	discontinuous sand and clay layers	"	?	"	"	"
HILL AFB SOIL PILE	mixture of sand and silty sand	"	?	permeable	?	?

? -- no information
GWT -- groundwater table

Extraction vents had to be installed below this clay lens to assure an effective extraction operation. Varying strata were also a concern at the gas station site in Bellview, Florida (Camp, Dresser, and McKee, 1987). Some layering of soil can make it easier to extract VOCs from soils where horizontal air channeling occurs through sand layers with subsequent VOC diffusion from less permeable layers.

The soil moisture content or degree of saturation is also important in that it is easier to draw air through drier soils. A case in point is that of the South Pacific Transportation site in Arizona where the soil was relatively dry (2 to 5% moisture content) (Johnson, 1988; Johnson and Sterrett, 1988). After seven months, 6500 kg of dichloropropene had been extracted using a moderate air flow rate of 85 to 250 cfm. Higher air flow rates tend to increase vapor removal because the radius of influence increases and more air is forced through the air filled pores. In addition, more air is pulled through the soil in a shorter time period.

Types and Magnitude of Contamination

The types and magnitude of chemical contamination encountered at the various sites are summarized in Table A-7. The common chemical contaminants extracted were trichloroethylene, 1,1,1-trichloroethane, methylene chloride, carbon tetrachloride, tetrachloroethylene, dichloroethylene, toluene, 1,3-dichloropropene, and gasoline constituents (benzene, toluene, ethylbenzene, and xylene). Most chemicals that have been successfully extracted have a low molecular weight and high volatility. Another common screening tool is the air-water partitioning coefficient, expressed in dimensionless terms as Henry's Law constant (See Table A-8). Most of the compounds have values of Henry's Law constants greater than 0.01. Vapor extraction has removed large quantities of volatile chemicals as demonstrated at several sites.

EXTRACTION SYSTEM OPERATION

At most sites, the initial VOC recovery rates were relatively high and then decreased asymptotically to zero with time (Oster and Wenck, 1988; Payne et al., 1985; Payne and Lisiecki, 1988; Terra Vac, 1987b). Vapor extraction is more effective at those sites where the more volatile chemicals are still present than when the spill is relatively recent. Several studies have indicated that intermittent venting from individual vents is probably more efficient in terms of mass of VOC extracted per unit of energy expended (Crow et al., 1987; Oster and Wenck, 1988; Payne and Lisiecki, 1988). This is especially true when extracting from soils where mass transfer is limited by the rate chemicals diffuse out of immobile air and water. Optimal operation of a soil vapor extraction system may involve taking individual vents in and out of service to allow time for liquid and gas diffusion and to change air flow patterns in the region being vented.

One of the major problems in the operation of a soil vapor extraction system is determining when the site is sufficiently clean to cease operation. Mass balances using initial and final soil borings have not been particularly successful in predicting the amount of chemical actually removed in a system (Anastos et al., 1985; Camp, Dresser, and McKee, 1988). Soil vapor

Table A-7. Pilot and Field Soil Vapor Extraction Systems --
Types and Magnitude of Contamination.

SITE	CHEMICALS IDENTIFIED	SPILL VOLUME	INITIAL CONTAMINATION LEVELS	FINAL CONTAMINATION LEVELS	AMOUNT EXTRACTED
FUEL MARKETING TERMINAL	gasoline hydrocarbons	>100000 gal	1.6 ft product on GWT 60-110 ppmv @ 16 ft, 3500-28000 @ 20 ft, 11000-51000 @ 21 ft	?	190 gallons
VALLEY MANUFACTURING	TCE, PCE, MC DCE, TCA	unknown	max concs: 2500mgTCE/kg 40 mgPCE/kg, 12 mgDCE/kg	being measured	being measured
INDUSTRIAL TANK FARM	carbon tetrachloride	200,000 lbs	70% of carbon tet contained in unsat. zone	? initial rate = 250 lb/day	>70% of spill volume
TIME OIL COMPANY	TCE, PCE, TTCA MC, TCA, DCE	unknown	from 5 ppm at 30 ft to over 1000 ppm at 6 in	current status unknown	240 lbs
SOLVENTS STORAGE TANK	TCA, TCE DCA, DCE	unknown	>10 mgTCA/m^3, 1 mgTCE/m^3	unknown, extraction rate decreased with time	11lb/day
TCAAP PILOT 1	TCE, TCA DCE, toluene + others	unknown	5 - 50 mgVOC/kg stained sediments 4-40 ft BLS TCE up to 8000 mg/kg	not determined	~1000 lbs
TCAAP PILOT 2	"	"	"	"	?
TCAAP SITE D	"	"	"	"	>84,000 lbs VOCs
TCAAP SITE G	TCE, TCA DCE, toluene + others	unknown	>1000 mgVOC/kg	not determined	>85,000 lbs VOCs
GAS STATION	gasoline	unknown	up to 10 in of gasoline on GWT, no HC @ 18 ft BLS	no free product 98% reduction of HC in GW	1200 lbs of gasoline
UNION 76 GAS STATION	benzene, toluene xylene, HCs	unknown	0.2 to 12.4 mgBTEX/kg highest conc. at 15 ft		22,000 lbs in 123 days
SOUTH PACIFIC RAILROAD	dichloropropene	150,000 lbs	30 to 60% of initial spill remaining in soil	less than 10 ppm in 40 soil samples	90,000 lbs
CUSTOM PRODUCTS	PCE	<5000 cu yd soil	8 to 5600 mgPCE/kg soil 92000 mg/m^3 in exhaust	17 ugPCE/kg soil after 280 d	62 - 76 kg in 35 days
ELECTRONIC MANUFACTURING	TCA chl. solvents	unknown	2000 ppmv organics in initial extracted gas	50 ppmv in exhaust (target is 20 ppmv)	>12000 lbs VOCs
PAINT STORAGE	acetone, ketones toluene, xylenes	over 400,000 cu yds soil	Total VOC in GW from 1 to 620,000 ug/L	Total VOC in GW from not detected to 10 ug/L	>7800 lbs after 165 days
THOMAS SOLVENT COMPANY	PCE, TCE, TCA	?	1700 lbs VOC in 1984	not determined	?
HILL AFB VERTICAL VENTS	jet fuel (JP4)	>25,000 gal total	up to 6200 mg/kg fuel in upper 5 ft of soil	system still in operational	1600 lbs in one-well vent test
HILL AFB LATERAL SYSTEM	"	"	200 - 900 mg/kg between 5 - 10 ft deep below detection below 50 ft deep		
HILL AFB SOIL PILE	"	"	soil vapor conc. up to 80000 ppb in top 10 ft		

BTEX -- benzene, toluene, ethylbenzene, and xylene
DCA -- dichloroethane
DCE -- dichloroethene
HC -- hydrocarbon
MC -- methylene chloride
PCE -- tetrachlorethene (perchloroethylene)

TCA -- trichloroethane
TCE -- trichloroethene
TTCA -- tetrachloroethane
VOC -- volatile organic chemical
na -- not applicable
ppmv -- parts per million by volume

Table A-8.   Dimensionless Henry's Law Constants for Typical
Organic Compounds.

Component	10°C	15°C	20°C	25°C	30°C
nonane	17.21519	20.97643	13.80119	16.92131	18.69235
n-hexane	10.24304	17.46626	36.70619	31.39026	62.70981
2-methylpentane	29.99747	29.35008	26.31372	33.72000	34.08841
cyclohexane	4.43291	5.32869	5.81978	7.23447	8.96429
chlorobenzene	0.10501	0.11884	0.14175	0.14714	0.19014
1,2-dichlorobenzene	0.07015	0.06048	0.06984	0.06417	0.09527
1,3-dichlorobenzene	0.09511	0.09769	0.12222	0.11649	0.16964
1,4-dichlorobenzene	0.09124	0.09177	0.10767	0.12957	0.15637
o-xylene	0.12266	0.15267	0.19704	0.19905	0.25164
p-xylene	0.18076	0.20427	0.26813	0.30409	0.37988
m-xylene	0.17689	0.20976	0.24859	0.30409	0.35656
propylbenzene	0.24446	0.30915	0.36623	0.44143	0.55072
ethylbenzene	0.14030	0.19073	0.24983	0.32208	0.42209
toluene	0.16397	0.20807	0.23071	0.26240	0.32480
benzene	0.14203	0.16409	0.18790	0.21581	0.28943
methyl ethylbenzene	0.15106	0.17762	0.20910	0.22807	0.30953
1,1-dichloroethane	0.15838	0.19200	0.23404	0.25545	0.31194
1,2-dichloroethane	0.05035	0.05498	0.06111	0.05763	0.06995
1,1,1-trichloroethane	0.41532	0.48635	0.60692	0.71119	0.84819
1,1,2-trichloroethane	0.01678	0.02664	0.03076	0.03719	0.05346
cis-1,2-dichloroethylene	0.11620	0.13787	0.14965	0.18556	0.23114
trans-1,2-dichloroethylene	0.25390	0.29815	0.35625	0.38625	0.48640
tetrachloroethylene	0.36410	0.46943	0.58614	0.69892	0.98487
trichloroethylene	0.23154	0.28208	0.35002	0.41690	0.51454
tetralin	0.03228	0.04441	0.05654	0.07643	0.10773
decalin	3.01266	3.53977	4.40641	4.78211	7.99952
vinyl chloride	0.64557	0.71049	0.90207	1.08313	1.12556
chloroethane	0.32666	0.40515	0.45727	0.49456	0.57484
hexachloroethane	0.25522	0.23641	0.24568	0.34129	0.41405
carbon tetrachloride	0.63696	0.80776	0.96442	1.20575	1.51951
1,3,5-trimethylbenzene	0.17344	0.19454	0.23736	0.27507	0.38711
ethylene dibromide	0.01291	0.02030	0.02536	0.02657	0.03216
1,1-dichloroethylene	0.66278	0.85851	0.90622	1.05860	1.27832
methylene chloride	0.06025	0.07147	0.10143	0.12098	0.14512
chloroform	0.07403	0.09854	0.13801	0.17207	0.22270
1,1,2,2-tetrachloroethane	0.01420	0.00846	0.03035	0.01022	0.02814
1,2-dichloropropane	0.05251	0.05329	0.07898	0.14592	0.11497
dibromochloromethane	0.01635	0.01903	0.04282	0.04823	0.06110
1,2,4-trichlorobenzene	0.05552	0.04441	0.07607	0.07848	0.11939
2,4-dimethylphenol	0.35678	0.28504	0.41986	0.20150	0.15074
1,1,2-trichlorotrifluoroethane	6.62785	9.09260	10.18462	13.03840	12.90375
methyl ethyl ketone	0.01205	0.01649	0.00790	0.00531	0.00442
methyl isobutyl ketone	0.02841	0.01565	0.01206	0.01594	0.02734
methyl cellosolve	1.89798	1.53517	4.82210	1.26297	1.53277
trichlorofluoromethane	2.30684	2.87580	3.34222	4.12815	4.90423

Adapted from Howe et al. (1986)

measurements in conjunction with soil boring and groundwater monitoring may be useful in determining the amount of chemical remaining in the soil. Risk analysis has been used to evaluate final clean up in at least one system (Ellgas and Marachi, 1988). Payne and Lisiecki (1988) suggest intermittent operation near the end of clean up. If there ceases to be a significant increase in vapor concentration upon restart, one can assume the site has been decontaminated.

Malot and Wood (1985) discuss use of in-situ soil air extraction in conjunction with groundwater pumping and treatment as a low-cost alternative for the clean up of petroleum and solvent spills. Large quantities of organic chemicals can be retained in the vadose zone by capillary forces, dissolution in soil water, volatilization, and sorption. If this product can be removed before it reaches the groundwater then the problem is mitigated. Since vapor transport is diffusion-controlled in the absence of air extraction, the vapor spreads horizontally, and a concentration gradient is established in the vertical direction as vapor diffuses back to the surface. Malot and Wood (1985) indicate that vapor extraction is effective in removing organic chemical vapor, sorbed chemical, and free product at the water table. This suggests that the soil should be decontaminated by vapor extraction before groundwater clean up can be completed. Vapor extraction becomes more cost-effective as the depth to groundwater increases, primarily because the cost of excavation becomes prohibitive.

The design and operation of soil vapor extraction systems can be quite flexible, allowing for changes to be made during the course of operation, with regard to vent placement or blower size, and air flows from individual vents. If the system is not operating effectively, changes in the vent placement or the capping the surface may improve it. At one site, the blowers were housed in modules with quick disconnect attachments. This allowed for portability, thus improving the removal efficiency by allowing for the blowers to be moved about the site to particular locations where extraction was required the most.

CONCLUSIONS

Based on the current state of the technology of soil vapor extraction systems, the following conclusions can be made:

1. Soil vapor extraction can be effectively used for removing a wide range of volatile chemicals over a wide range of conditions.

2. The design and operation of these systems is flexible enough to allow for rapid changes in operation, thus, optimizing the removal of chemicals.

3. Intermittent blower operation is probably more efficient in terms of removing the most chemicals with the least energy, especially in systems where chemical transport is limited by diffusion through air or water.

4. Volatile chemicals can be extracted from clays and silts but at a slower rate. Intermittent operation is certainly more efficient

under these conditions.

5. Air injection and capping a site have the advantage of controlling air movement, but injection systems need to be carefully designed.

6. Extraction vents are usually screened from a depth of from 5 to 10 below the surface to the groundwater table. For thick zones of unsaturation, maximum screen lengths of 20 to 30 feet are specified.

7. Air/water separators are simple to construct and should probably be installed in every system.

8. Installation of a cap over the area to be vented reduces the chance of extracting water and extends the path that air follows from the ground surface, thereby increasing the volume of soil treated.

9. Incremental installation of vents, while probably more expensive, allows for a greater degree of freedom in design. Modular construction, where the most contaminated zones are vented first, is preferable.

10. Use of soil vapor probes in conjunction with soil borings to assess final clean up is less expensive than use of soil borings alone. It is usually impossible to do a complete materials balance on a given site because most sites have an unknown amount of VOC on the soil and in the groundwater.

11. Soil vapor extraction systems are usually only part of a site remediation system.

12. While a number of variables intuitively affect the rate of chemical extraction, no extensive study to correlate variables to extraction rates has been identified.

REFERENCES

Alliance Technologies Corp. 1987. Quality Assurance Project Plan, Terra Vac Inc., In-Situ Vacuum Extraction Technology, SITE Demonstration Project, Valley Manufactured Products Site, Groveland, MA. Contract No. 68-03-3255. Bedford, MA. September 1, 1987.

American Petroleum Institute. 1984. Forced Venting To Remove Gasoline Vapor From a Large-Scale Model Aquifer. American Petroleum Institute. Environmental Affairs Department. Washington, D.C. June, 1984.

American Petroleum Institute. 1985. Subsurface Venting Of Hydrocarbon Vapors From an Underground Aquifer. API Publication No. 4410. Health and Environmental Sciences Department. Washington, D.C. September, 1985.

Anastos, G.J., P.J. Marks, M.H. Corbin, and M.F. Coia. 1985. Task 11. In Situ Air Stripping of Soils, Pilot Study, Final Report. Report No. AMXTH-TE-TR-85026. U.S. Army Toxic & Hazardous Material Agency. Aberdeen Proving Grounds. Edgewood, MD. 88 pp. October 1985.

AWARE, Inc. 1987. Phase I - Zone I Soil Decontamination Through In-Situ Vapor Stripping Processes, Final Report. Contract Number 68-02-4446. U.S. Environmental Protection Agency, Small Business Innovative Research Program, Washington, D.C. April 1987.

Bennedsen, M.B. 1987. Vacuum VOC's from Soil. Pollution Engineering, 19(2):66-68.

Bennedsen, M.B., J.P. Scott, and J.D. Hartley. 1985. Use of Vapor Extraction Systems for In Situ Removal of Volatile Organic Compounds from Soil. Proceedings of National Conference on Hazardous Wastes and Hazardous Materials, HMCRI, pp. 92-95.

Camp Dresser and McKee, Inc. 1987. Interim Report For Field Evaluation of Terra Vac Corrective Action Technology at a Florida Lust Site. Contract No. 68-03-3409. U.S. Environmental Protection Agency. Edison, NJ. December 21, 1987.

CH$_2$M Hill Inc. Remedial Planning/Field Investigation Team. 1985. Verona Well Field - Thomas Solvent Company, Battle Creek, Michigan, Operable Unit Feasibility Study. Contract No. 68-01-6692. U.S. Environmental Protection Agency. Chicago, IL. June 17, 1985.

CH$_2$M Hill Inc. 1987a. Operable Unit Remedial Action, Soil Vapor Extraction at Thomas Solvents Raymond Road Facility, Battle Creek, MI, Quality Assurance Project Plan. U.S. Environmental Protection Agency. Chicago, IL. August 1987.

CH$_2$M Hill Inc.   1987b.   Appendix B - Sampling Plan, Operable Unit Remedial
    Action; Creek, MI. U.S.E.P.A.   Chicago, IL.   October 1987.

Connor, R.   1988.   Case Study of Soil Venting.   Pollution   Engineering.
    20(7):74-78.

Crow, W.L., E.P. Anderson, and E.M. Minugh.   1987.   Subsurface Venting of
    Vapors Emanating from Hydrocarbon Product on Ground Water.   Ground Water
    Monitoring Review.   7(1):51-57.

Dynamac Corporation.   1986.   Literature Review of Forced Air Venting to Remove
    Subsurface Organic Vapors from Aquifers and Soil.   Subtask Statement No.
    3. U.S. Air Force Engineering and Services Center, Tyndall AFB, FL, 30
    pp., July 28, 1986.

Ellgas, R.A. and N.D. Marachi.   1988.   Vacuum Extraction of Trichloroethylene
    and Fate Assessment in Soils and Groundwater: Case Study in California.
    Proceedings of the 1988 Joint CSCE-ASCE National Conference on
    Environmental Engineering.   Vancouver, B.C., Canada.   pp. 794-801.   July
    13-15, 1988.

Enviresponse, Inc.   1987.   Demonstration Test Plan, In-Situ Vacuum Extraction
    Technology, Terra Vac Inc., SITE Program, Groveland Wells Superfund
    Site, Groveland, MA.   EERU Contract No. 68-03-3255, Work Assignment 1-
    R18, Enviresponse No. 3-70-06340098.   Edison, NJ. November 20, 1987.

Howe, G.B., M.E. Mullins, and T.N. Rogers.   1986.   Evaluation and Prediction
    of Henry's Law Constants and Aqueous Solubilities for Solvents and
    Hydrocarbon Fuel Components Volume I: Technical Discussion.   USAFESC
    Report No. ESL-86-66. U.S. Air Force Engineering and Services Center,
    Tyndall AFB, FL.   86 pp.

Hutzler, N.J., J.S. Gierke, and L.C. Krause.   1988.   Movement of Volatile
    Organic Chemicals in Soils.   In: Reactions and Movement of Organic
    Chemicals in Soils, B.L. Sawhney, ed., Soil Science Society of America,
    Madison, WI, in press.

Johnson, J.J.   1988.   In-Situ Soil Air Stripping: Analysis of Data From A
    Project Near Benson, Arizona.   M.S. Thesis, Colorado School of Mines,
    Golden, CO.   March 16, 1988.

Johnson, J.J. and R.J. Sterrett.   1988.   Analysis of In Situ Soil Air
    Stripping Data.   Proceedings of the 5th National Conference on Hazardous
    Wastes and Hazardous Materials.   HMCRI.   Las Vegas, NV.   April 19-21,
    1988.

Malot, J.J., and P.R. Wood.   1985.   Low Cost, Site Specific, Total Approach to
    Decontamination.   Conference on Environmental and Public Health Effects
    of Soils Contaminated with Petroleum Products, University of
    Massachusetts, Amherst, MA.   October 30-31, 1985.

Malot, J.J., J.C. Agrelot, and M.J. Visser. 1985. Vacuum: Defense System for
    Ground Water VOC Contamination. Fifth National Symposium on Aquifer
    Restoration and Ground Water Monitoring, Columbus, Ohio. May 21-24,
    1985.

Malot, J.J. 1985. Unsaturated Zone Monitoring and Recovery of Underground
    Contamination. Fifth National Symposium on Aquifer Restoration and
    Ground Water Monitoring, Columbus, Ohio. May 21-24, 1985.

Markley, D.E. 1988. Cost Effective Investigation and Remediation of
    Volatile-Organic Contaminated Sites. Proceedings of the 5th National
    Conference on Hazardous Wastes and Hazardous Materials, HMCRI, Las
    Vegas, NV. April 19-21, 1988.

Oak Ridge National Laboratory. 1987. Draft: Preliminary Test Plan, In-Situ
    Soil Venting Demonstration, Hill AFB, Utah. U.S. Air Force Engineering
    and Services Center, Tyndall AFB, FL.

Oster, C.C. and N.C. Wenck. 1988. Vacuum Extraction of Volatile Organics
    from Soils. Proceedings of the 1988 Joint CSCE-ASCE National Conference
    on Environmental Engineering, Vancouver, B.C., Canada. pp. 809-817.
    July 13-15, 1988.

Payne, F.C., C.P. Cubbage, G.L. Kilmer, and L.H. Fish. 1986. In Situ Removal
    of Purgeable Organic Compounds from Vadose Zone Soils. Purdue
    Industrial Waste Conference. May 14, 1986.

Payne, F.C. and J.B. Lisiecki. 1988. Enhanced Volatilization for Removal of
    Hazardous Waste from Soil. Proceedings of the 5th National Conference
    on Hazardous Wastes and Hazardous Materials, HMCRI, Las Vegas, NV.
    April 19-21, 1988.

Radian Corp. 1987. Installation Restoration Program Phase II Draft Report.
    U.S. Air Force, Hill AFB, UT. July 1987.

Rollins, Brown, and Gunnell, Inc. 1985. Subsurface Investigation and
    Remedial Action, Hill AFB JP-4 Fuel Spill, Provo, Utah. U.S. Air Force,
    Hill AFB, UT. December 1985.

Terra Vac, Inc. 1987. Demonstration Test Plan In-Situ Vacuum Extraction
    Technology. Enviresponse No. 3-70-06340098. Terra Vac, Inc., SITE
    Program, Groveland Wells Superfund Site, Groveland, MA. November, 1987.

Terra Vac, Inc. 1987. Union 76 Gas Station Clean-up, Bellview, Florida.
    Florida Department of Environmental Regulation, Tallahassee, FL.

Texas Research Institute. 1986. Examination of Venting For Removal of
    Gasoline Vapors From Contaminated Soil. American Petroleum Institute,
    March, 1980, (Reprinted in 1986).

Thornton, S.J. and W.L. Wootan. 1982. Venting for the Removal of Hydrocarbon
    Vapors from Gasoline Contaminated Soil. J. Environmental Science and

Health, A17(1):31-44.

Thornton, S.J., R.E. Montgomery, T. Voynick, and W.L. Wootan. 1984. Removal
of Gasoline Vapor from Aquifers by Forced Venting. Proceedings of the
1984 Hazardous Materials Spills Conference, Nashville, TN. pp. 279-286.
April 1984.

Towers, D.S., M.J. Dent, and D.G. Van Arnam. 1988. Evaluation of In Situ
Technologies for VHOs Contaminated Soil. Proceedings of the 5th
National Conference on Hazardous Wastes and Hazardous Materials, HMCRI,
Las Vegas, NV. April 19-21, 1988.

Treweek, G.P. and J. Wogec. 1988. Soil Remediation by Air/Steam Stripping.
Proceedings of the 5th National Conference on Hazardous Wastes and
Hazardous Materials, HMCRI, Las Vegas, NV. April 19-21, 1988.

U.S. Army. 1986a. "Twin Cities Army Ammunition Plant In-Situ Volatilization
System, Site G, First Week Operations Report", Twin Cities Army
Ammunition Plant, New Brighton, MN, March 1986.

U.S. Army. 1986b. "Twin Cities Army Ammunition Plant In-Situ Volatilization
System Site D, Operations Report," Twin Cities Army Ammunition Plant,
New Brighton, MN, September 8, 1986.

U.S. Army. 1987a. "Twin Cities Army Ammunition Plant In-Situ Volatilization
System Site D Operations Report," Twin Cities Army Ammunition Plant, New
Brighton, MN, September 1, 1987.

U.S. Army. 1987b. "Twin Cities Army Ammunition Plant In-Situ Volatilization
System Brighton, MN, October 2, 1987.

U.S. Army. 1987c. "Twin Cities Army Ammunition Plant In-Situ Volatilization
System Site G, Emissions Control System Operations Report," Twin Cities
Army Ammunition Plant, New Brighton, MN, September 1, 1987.

U.S. Army. 1987d. "Twin Cities Army Ammunition Plant In-Situ Volatilization
System Site G, Emissions Control System Operations Report," Twin Cities
Army Ammunition Plant, New Brighton, MN, October 2, 1987.

Wenck Associates, Inc. 1985. "Project Documentation: Work Plan, ISV/In-Situ
Volatilization, Sites D and G, Twin Cities Army Ammunition Plant,"
Federal Cartridge Corporation, New Brighton, MN. September 1985.

Weston, Roy F., Inc. 1985. Appendices -- Task 11, In-Situ Solvent Stripping
From Soils Pilot Study. Installation Restoration General Environmental
Technology Development Contract DAAK11-82-C-0017. U.S. Army Toxic and
Hazardous Materials Agency, Aberdeen Proving Grounds, MD. May 1985.

Woodward-Clyde Consultants. 1985. Performance Evaluation Pilot Scale
Installation and Operation Soil Gas Vapor Extraction System Time Oil
Company Site Tacoma, Washington, South Tacoma Channel, Well 12A Project.
Work Assignment No. 74-ON14.1, Walnut Creek, CA. December 1985.

Wootan, W.L. and T. Voynick. 1984. Forced Venting to Remove Gasoline Vapor
     from a Large-Scale Model Aquifer. Texas Research Institute, Inc., Final
     Report to American Petroleum Institute.

ACKNOWLEDGEMENTS

This paper is based on work supported by the United States Environmental
Protection Agency under assistance agreement CR-814319-01-1. This paper has
not been peer-reviewed. Any opinions, findings, and conclusions are those of
the authors and do not necessarily reflect the views of the Environmental
Protection Agency. Mention of trade names or commercial products does not
constitute and endorsement or recommendation for use.

Appendix B
Applicability and Limits of Soil Vapor Extraction
for Sites Contaminated with Volatile Organic Compounds

Joseph Danko, P.E.[a]

INTRODUCTION

Many communities throughout the United States and the rest of the world have discovered that their soil and groundwater are contaminated with volatile organic compounds (VOCs). This contamination results from activities such as poor disposal practices; careless handling of VOCs at transfer and storage facilities that results in surface spillage; and leaking underground storage tanks (USTs).

Soil vapor extraction (SVE) is an in situ technology currently used for the removal of VOCs from vadose zone soils. The use of this technology has grown significantly over the last 5 years because it generally costs less than other alternatives (especially excavation with subsequent disposal or treatment), is easy to implement, and has the potential for significant removal rates. However, many criteria need to be evaluated before selecting this technology for a VOC site. Without this evaluation, ineffective or noncompliant cleanup goals could result. These criteria include identifying the contaminants and their characteristics; defining the nature, extent, and volume of contamination; investigating in detail the vadose zone characteristics affecting vapor transport; determining the depth to and nature of the underlying saturated zone; and evaluating air emissions requirements and vadose zone cleanup standards.

The purpose of this paper is to provide a list of criteria to evaluate the applicability of soil vapor extraction at a potential site. Wherever possible, rules of thumb and limitations are provided to assist the user in this assessment.

SVE DESCRIPTION

Soil vapor extraction is a physical means of removing VOCs from contaminated soil. The typical SVE system consists of a network of vacuum extraction wells screened in the contaminated zone. The extraction wells are joined together by a common header pipe, which is connected to a vapor water

---

[a]CH₂M Hill, P.O. Box 428, Corvalis, OR 97339

163

separator where water is removed. The separator is then connected to a positive displacement blower, which provides a negative pressure gradient in the subsurface. Discharge from the blower is vented to the atmosphere or connected to an offgas treatment system, depending upon air emissions requirements and the nature and extent of VOC contamination.

The subsurface vacuum created by the blower pulls VOC-laden vapors through the subsurface into the extraction wells. Pulling air through soil voids disrupts the equilibrium concentration between liquid or sorbed contaminants and VOCs in the gas phase. A concentration gradient is established from liquid or sorbed contaminants in soil interstices and micropores and VOCs present in the gas phase. Evaporation of contaminants to the gas phase occurs in the same manner in which air stripping removes contaminants from groundwater. The vacuum also decreases pressure in soil voids, thereby causing the release of additional VOCs.

ADVANTAGES OF SVE

The current increase in the use of SVE stems from the advantages, including ones related to regulatory factors, associated with the in situ nature of the technology. These advantages include the following:

- SVE is minimally intrusive to contaminated soils. During construction and operation, the potential release of VOCs to onsite and offsite receptors is insignificant when compared to excavation and removal of contaminated soils.

- SVE is not a complicated technology to implement. As described earlier, the typical system contains extraction wells, piping, positive displacement blowers, and standard instrumentation and controls. However, optimal sizing and operation of this equipment on a medium-sized or larger site does require assistance from experienced personnel. Also, if flammable VOCs are present, care must be taken to avoid fire and/or explosion in the SVE system.

- The use of SVE can result in a relatively quick reduction in VOCs. In addition, full-scale SVE systems have been successful at many sites across the country, and the systems can be constructed safely with proper instrumentation and controls.

- Vadose zone VOC contamination often acts as a source input to groundwater contamination. SVE can reduce or effectively eliminate the vadose zone source input, thereby drastically decreasing the time required for saturated zone pump-and-treat alternatives.

- SVE, when applicable, is more cost effective than other in situ technologies. When compared to excavation costs (with subsequent disposal or treatment), its costs can easily be an order of magnitude lower.

- With the land ban on solvent tainted soils, SVE offers a viable alternative technology to excavation with disposal or treatment.

APPLICABILITY AND LIMITATIONS OF SVE

With all of the advantages of this technology, it is easy to understand that its use is on the rise in site cleanups. However, there are many criteria that need to be evaluated before assessing the limitations and applicability of this technology for a given site. The following text describes criteria to be evaluated when considering this technology for site remediation.

Volatility of Contaminants

SVE is applicable if VOCs are the primary contaminants in the soil. As a guideline, a compound is a likely candidate if it has both of these characteristics:

- Vapor pressure (P*) of 1.0 mm or more of mercury at 20°C

- Henry's Law constant greater then 100 atmospheres/mole fraction (in the moderate range), or dimensionless Henry's Law constants greater than 0.01

Examples of VOCs amenable to SVE are listed in Table B-1.

TABLE B-1.   VOCs AMENABLE TO SVE

Compound	Henry's Law Constant [a] /P*[b]
1,1,1-Trichloroethane	1,535/100
Trichloroethylene	590/60
Tetrachloroethylene	1,261/14
1,1-Dichloroethylene	1,086/500
1,2-Dichloroethane	60/61
Benzene	302/76
Toluene	367/22

[a]Henry's Law = atm./mole fraction
[b]P* = mm of Hg at 20°C

Compounds that are more difficult to extract would be trichlorobenzene and diesel and other large molecular weight petroleum fuels.

Nature, Volume, and Extent of Contamination

SVE is an especially attractive option if the contamination is beneath a building or surrounding the support structure of a building. In this case, excavation could unearth the existing support structure of the building or extensive demolition of the building would be required before excavating the contaminated soil.

The quantity of contaminants spilled is important to quantify through investigations of site history and site operations and by conducting soil gas and soil boring analyses. If the total quantity of contaminants is low and there is no apparent exposure pathway or regulatory requirement to expedite soil cleanup, it may be more cost effective to leave the contaminated soil in place and allow natural transport and degradation processes to govern the fate of contaminants.

Another viable alternative for a point source of contamination may be excavation with disposal and treatment, depending upon the type and concentration of contaminants present.

Groundwater Depth

SVE usually is more effective than other alternatives if groundwater depth is greater than 20 feet and vadose zone contamination extends to the groundwater table. If groundwater is less than 5 to 10 feet, then soil washing or excavation without SVE could be more cost effective, depending on:

- Quantity to be excavated for disposal or treatment (smaller quantities would favor excavation);

- Whether state and federal applicable or relevant and appropriate requirements (ARARs) would allow excavation and disposal or treatment.

In addition, SVE usually is a good alternative for sites where the majority of contamination is in the vadose zone. Sites where the majority of contaminants are in the saturated zone must be evaluated to compare dewatering and subsequent SVE with groundwater pump and treat in conjunction with SVE. This evaluation should consider:

- Saturated zone characteristics

- Overall cleanup schedule

- Regulations governing discharge.

Characteristics of Contaminated Soil

SVE is typically more applicable to cases where the contaminated unsaturated zone is relatively permeable (hydraulic conductivities in excess of 10^{-3} or 10^{-2} cm^3/cm^2 sec) and uniform. Sands and gravels are especially amenable to SVE. However, the technology has been used in less permeable clayey or silty soils with some success. Agrelot et al., 1985 and Applegate et al, 1988, have demonstrated removal of contaminants in soils with conductivities ranging from 10^{-3} to 10^{-6}. This success could be due to the presence of more permeable sand and gravel strata typically found in alluvial settings or the relatively low moisture contents in the finer-grained soils (Bennedsen, 1987).

Michaels and Stinson (1989) have found that porosity appears to be a more important characteristic to consider when evaluating the applicability of

SVE. These conclusions are based on the results of the SITE program demonstration test of Terra Vac's vacuum extraction system in Groveland, Massachusetts. Significant VOC removal rates were achieved in relatively impermeable clays (hydraulic conductivities of 10^{-8} cm^3/cm^2-sec) and more permeable sands (hydraulic conductivities of 10^{-4} cm^3/cm^2-sec). Both soil strata had porosities between 40 and 50 percent.

It is important to note that many other physical characteristics that will influence vapor transport must be investigated at the site before a system can be designed and constructed. These characteristics include stratigraphy, particle size distribution, moisture content, bulk density, and particle density. Such parameters are currently used by SVE vendors and other design professionals in flow models, and empirically to provide additional information on the airflow and vacuum requirements of the system.

Emission Controls

More and more states are now requiring SVE offgas treatment to comply with stringent ambient air quality criteria, rather than allowing direct discharge to the atmosphere. Offgas treatment methods include:

- Vapor phase GAC with offsite or onsite regeneration or offsite disposal

- Condensation

- Incineration or flaring.

The need for offgas treatment can be determined by investigating state and federal air emissions requirements. If offgas treatment is required, the most cost-effective technology needs to be determined by evaluating the type, concentration and quantity of contaminants; permit requirements; and availability of required utilities. If expensive and complex offgas treatment systems are needed, the SVE is less favorable than other alternatives. It is critical that estimates of total costs for the SVE system include offgas emission requirements, if any, to allow a valid comparison with other in situ alternatives and with the excavation alternative.

Schedule for Cleanup

Soil vapor extraction is neither the fastest nor the slowest method of VOC removal from the vadose zone. The Table B-2 lists a range of vadose zone remedial alternatives and qualitative estimates of cost and cleanup rates.

The selection of an alternative depends upon the relative magnitude of the cost difference for the specific site, and the impact of immediate removal (excavation) vs. slower removal rates. Soil vapor extraction is more applicable when site conditions and regulatory requirements can accommodate a removal rate taking weeks or months at moderately contaminated sites, to several months or years at heavily contaminated sites.

TABLE B-2.   VADOSE ZONE REMEDIAL ALTERNATIVES

Alternatives	Cost	Cleanup Rate
Natural Attenuation	Minimal	Slow
Soil Washing	Low to Moderate	Slow to Moderate
SVE	Low	Moderate
Excavation	High	Fast

Vadose Zone Cleanup Standards

The site's vadose zone cleanup standard performance objective (PO) has a significant bearing on the applicability of SVE. The following text describes two aspects of the vadose zone cleanup standard--concentration limits and PO testing--that need to be considered.

Concentration limits--

Soil vapor extraction is not as applicable if the concentration limits are low compound-specific limits (eg., 5 ug/kg tetrachloroethylene or 10 ug/kg trichloroethylene) to be achieved in a short duration of time. The performance of SVE at such low levels has not been widely demonstrated, especially in nonhomogeneous soils.

However, the site is a good candidate for SVE if the concentration limits to be achieved are high ug/kg limits (e.g., total VOCs greater than 500 ug/kg), or if the PO is vapor analysis from the extraction well system (e.g., 750 ug/l tetrachloroethylene).

PO Testing--

Soil vapor extraction is not as applicable if the PO is to be verified by a statistical soil sampling grid with low compound-specific concentrations. Less permeable hot spots could remain at higher concentrations for a significant amount of time beyond the time required to remove the majority of contaminants. However, SVE would be a likely candidate if the PO is to be measured by vapor stream analysis from the combined extracted vapor.

This criterion cannot be overlooked when considering the applicability of SVE at a site. If the site is a good candidate after considering all of the factors discussed above, one could implement the technology and remove 95% or more of the contaminants present. This removal probably would be a significant environmental benefit, but the responsible party could be left with a liability if low compound-specific criteria are not obtained. This situation is highly probable given the low cleanup concentration limits in some states, and is a predicament none of us want to experience.

SUMMARY

Soil vapor extraction has been demonstrated as an effective technology for the removal of VOCs from soils contaminated from leaking USTs and surface spills. There are many advantages to the use of SVE due to the in situ nature of the technology, the effective removal of VOCs, and the ease of implementation and operation. However, many criteria must be evaluated before selecting SVE for a VOC-contaminated site. Failure to evaluate these criteria before selecting the technology could result in ineffective or noncompliant cleanup goals.

REFERENCES

Agrelot, Jose C., James J. Malot, and Melvin J. Visser. Vacuum Defense System for Groundwater VOC Contamination. Presented at the Fifth National Symposium on Aquifer Restoration and Ground Water Monitoring, Columbus, Ohio, May 21-24, 1985

Applegate, Joseph, John K. Gentry, and James J. Malot. Vacuum Extraction of Hydrocarbons from Subsurface Soils at a Gasoline Contamination Site. 1988.

Bennedsen, M.B. et al. Use of Vapor Extraction Systems for In Situ Removal of Volatile Organic Compounds from Soil. 1987.

Michaels, P.A. and Stinson, M.K. Technology Evaluation Report. SITE Program Demonstration Test Terra Vac In Situ Vacuum Extraction System Groveland, Massachusetts. Contract No. 68-03-3255. 1989.

Appendix C
Soil Gas Surveys in Support of
Design of Vapor Extraction Systems

H. B. Kerfoot[a]

ABSTRACT

Soil-gas surveys have proven to be a useful tool in preliminary
delineation of subsurface contamination by volatile organic compounds.
Delineation of soil-gas concentrations is also frequently used to choose the
locations of vapor extraction systems for cleanup of vadose zone contamination by volatile organic compounds. The subsurface fate and transport
of these compounds can drastically affect static soil-gas concentrations of
them and must be considered in evaluation of soil-gas survey results. In
addition, static soil-gas conditions can be different than those expected
during operation of a vapor extraction system. However, within these con-
straints, soil-gas surveys can provide valuable data in support of design and
installation of vapor extraction systems for vadose-zone cleanup. At a site
in California, an areal soil-gas survey was used to choose the locations of
vapor-extraction wells, and vertical borings with soil-gas depth profiles were
used to select the withdrawal depth and to install monitoring devices and the
actual vapor wells.

INTRODUCTION

Soil-gas sampling and analysis is a technique that has recently found
many applications in responses to soil and ground-water contamination.
Although soil-gas sampling techniques have been available since the turn of
the century, the technology has only recently been applied to subsurface
contamination. However, due to the cost-effectiveness and rapidity of the
technology relative to traditional soil or ground-water sampling and analysis,
its use has grown rapidly. Although the technology is limited to volatile
analytes, those are frequent problem compounds. Table C-1 lists the 25 most
frequently encountered ground-water contaminants at CERCLA (Superfund) sites
from a CERCLA Section 301 list. It can be noted that the majority of them are
volatile and thus are amenable to application of this technology for detection
and delineation of contamination. In addition to these compounds, gasoline
and jet fuels are quite volatile. The high volatility of these compounds
makes the use of soil vapor extraction an attractive potential alternative to
soil removal for cleanup. In fact, a soil-gas survey to delineate the extent
of contamination is frequently an initial step in design of soil vapor

---

[a]Kerfoot and Associates, 3057 Hebard Drive, Las Vegas, NV 89121

Table C-1. Most Frequently Reported Groundwater
Contaminants at 546 Superfund Sites.

Rank	Compound	Percent
1	Trichloroethene(TCE)[b]	33
2	Lead	30
3	Toluene[b]	28
4	Benzene[b]	26
5	Polychlorinated Biphenyls(PCBs)	22
6	Chloroform[b]	20
7	Tetrachloroethene(PCE)[b]	16
8	Phenol	15
9	Arsenic	15
10	Cadmium	15
11	Chromium	15
12	1,1,1-Trichloroethane[b]	14
13	Zinc	14
14	Ethylbenzene[b]	13
15	Xylene	13
16	Methylene Chloride	12
17	trans-1,2-Dichloroethene[b]	11
18	Mercury	10
19	Copper	9
20	Soluble Cyanide Salts	8
21	Vinyl Chloride[b]	8
22	1,2-Dichloroethane[b]	8
23	Chlorobenzene[b]	8
24	1,1-Dichloroethane[b]	8
25	Carbon Tetrachloride[b]	7

[a]Source: CERCLA 301 C Study
[b]Compounds amenable to detection by soil-gas analysis

extraction system. However, both soil-gas surveys and soil vapor extraction are also subject to influences from vadose-zone properties that can hamper their effectiveness. In this paper, factors are discussed that can influence soil-gas survey data and thus parameters used in the design of soil vapor extraction systems, cleanup criteria, and monitoring of the cleanup process. Two case studies are presented; one is intended to demonstrate the effect of subsurface transformations on soil-gas surveys to delineate fuel contamination, while the other presents a soil-gas survey that was used to provide real-time data for installation of a soil vapor extraction system and a monitoring system. Although soil-gas surveys provide information about a site that describes it without bulk gas flow, which is a fundamental aspect of a soil-venting system, data from a soil-gas survey can be interpreted to provide information to be used in design of an *in situ* soil-stripping system.

BASIS OF THE TECHNOLOGY

Soil-gas survey technology is based on transport of volatile compounds from contaminated soil or water through the vadose zone to the atmosphere. This mass-transport can be described by a three-step model. The first step in the mass transport is phase transfer from the condensed phase (soil or water) to the gas phase. Once in the gas phase the contaminant travels through the vadose zone in response to a concentration gradient. Once the contaminant reaches the atmosphere it is rapidly carried away. Under relatively static subsurface conditions (e.g., in the absence of bulk gas flow), contaminant transport through the vadose zone is by diffusion. Only at shallow depths (<0.3m) with uncovered, high-porosity soils does wind have an effect.

Subsurface Diffusion

Diffusion of gases in soils has been studied since the early 1900's by soil scientists in the study of oxygen, nitrogen, and carbon dioxide transport and respiration. In addition, diffusion in porous media is a subject of considerable interest to chemical engineers in modeling the actions of catalysts in pellet form. Diffusion can be described by Fick's Law:

$$dm/dt = -D(d^2C/dz^2) \tag{1}$$

where dm/dt is the mass flux through a unit area, D is the diffusion coefficient (cm^2/sec), C is concentration and z is distance. Under steady-state conditions;

$$dC/dz = \text{constant} \tag{2}$$

indicating a linear concentration profile, increasing towards the vapor source.

Typically, the soil-gas diffusion coefficient, D_s, is expressed as the diffusion coefficient for the compound in air, D_a, multiplied by a correction for the fraction of the volume occupied by soil, water and solids and a factor to correct for the tortuosity of the diffusion path. Table C-2 lists several published tortuosity relationships between D_s and D_a. It can be noted that all of the relations incorporate the air-filled porosity, D_a, to at least the first power. Since the correction for the solid-filled volume also includes the air-filled porosity, these relations point out the sensitivity of soil-gas transport to that parameter. The data in Table C-2 are presented only to demonstrate the significant dependence of soil-gas transport on f_a, to point out the fact that minimal soil-gas diffusion occurs at air-filled porosities below 5 to 10 percent, and to indicate the inconsistent and often non-linear relationship between D_a and D_s. The first two of these points can be useful in evaluating the applicability of soil-gas-survey technology at a given site, and the last point provides insight into one cause of uncertainty in modelling the risk to ground-water quality due to gas-phase contaminant migration from soil to ground water. Such risk assessments can be an important factor in setting vadose-zone cleanup criteria.

The steady-state approximation for soil-gas diffusion can be useful as a limiting situation and agrees with observations in several instances.

TABLE C-2.   TORTUOSITY RELATIONSHIPS

Author	D_s/D_a*	Comments
Blake and Paige	$0.62f_a$ to $0.80f_a$	D_s) 0 when f_a <10%
Van Bavel	$0.61f_a$	
Marshall	$f_a^{3/2}$	
Millington	$(f_a/f)^2 f_a^{4/3}$	
Wesseling	$0.9f_a$ - 0.1	D_s) 0 at f_a ~ 11%
Grable and Siemer		D_s/D_a ~ 0.02 when f_a ~ 10%
Lemon and Erickson		D_s/D_a ~ 0.005 when f_a ~ 5%

* f_a - air-filled porosity; f - total porosity

However, shallower water tables, greater water infiltration, higher degrees of saturation, and higher organic carbon content of soils may hamper the achievement of steady state.

Surface or subsurface barriers to soil-gas transport can drastically affect soil-gas concentrations above and below them. From the discussion above, it can be seen that barriers to soil-gas transport are zones with air-filled porosity below 5 to 10 percent. Such zones would include saturated clay lenses, perched water bodies, pavement, and buildings. Standard mass-transport considerations indicate that such barriers will elevate soil-gas concentrations on one side (upstream in the mass flow) of them and will depress them on the other. Such barriers can also promote horizontal diffusion of the contaminants.

SUBSURFACE TRANSFORMATIONS

Organic contaminants can undergo a variety of subsurface transformations, both microbiologically catalyzed and chemical. Fuels are composed of reduced organic compounds and present a potential source of energy for subsurface microorganisms. Through oxidation of these compounds in several steps by several microorganisms, the organic carbon in the fuels becomes mineralized to inorganic carbon or carbon dioxide. Because of the high energy yield from these oxidations, microorganisms can use energy not just for respiration but for reproduction. The resulting microbial "population explosion" can obliterate hydrocarbons from soil gases when sufficient water, other nutrients, and electron acceptors are present. Workers at the U.S. EPA lab in

Ada, Oklahoma, have observed subsurface biooxidation of creosote that was
limited only by the rate of oxygen diffusion from the atmosphere to the
creosote. In zones of residual fuel saturation, such biodegradation can be
inhibited because of the lack of water and the toxicity of the fuels to the
microorganisms responsible for it. However, when the supply of oxygen is
increased, such as would be the case in a soil-venting effort, microorganisms
in nearby areas with sufficient water can be stimulated to increased levels of
mineralization of these compounds over that which takes place under static gas
mass conditions. Work to investigate this issue by the U.S. Navy Civil
Engineering Laboratory in Port Hueneme, California, is currently underway.

The impact of the potential for subsurface biodegradation of fuel
hydrocarbons on soil-gas surveys is to create a significant potential for
false negative results, where soil-gas hydrocarbons are not detected above
ground-water contamination or near shallow soil contamination. Figure C-1
shows the total hydrocarbon concentrations at a site with four 23-foot
diameter waste jet fuel tanks. Ground water is encountered at 80 to 90 feet
and the regional ground-water flow direction is to the southeast, and three
monitoring wells were installed during the soil-gas survey in hopes of
obtaining a RCRA permit to close the site. Table C-3 lists the analytical
results from those three wells. It should be noted that one of the downgradient wells had 8 feet of jet fuel present, and Table C-3 shows analytical
results for the ground water and the fuel layer.

TABLE C-3. MONITORING WELL DATA (mg/L).

WELL #	ETHYLBENZENE	XYLENES	BENZENE	TOLUENE
11	0.3	ND	ND	ND
13	ND	ND	ND	ND
12	6000	9200	440	1510
12 (8' NAPL)	520	4000	820	900

Because it was apparent that the contamination could well extend past the
extent indicated by the soil-gas hydrocarbon concentrations, and
biodegradation could be creating false negative results, a soil-gas survey
measuring soil-gas carbon dioxide concentrations was performed downgradient.
Figure C-2 shows the results of that survey. These data indicate a
significant soil-gas carbon dioxide anomaly downgradient of the well with pure
product in it. When a monitoring well was installed in that anomaly, a 4-foot
layer of fuel was encountered.

These results are presented only to demonstrate the high degree of
variability in soil-gas hydrocarbon behavior encountered at the same site.
Near the source (leaky tanks), floating fuel was adequately indicated by soil-

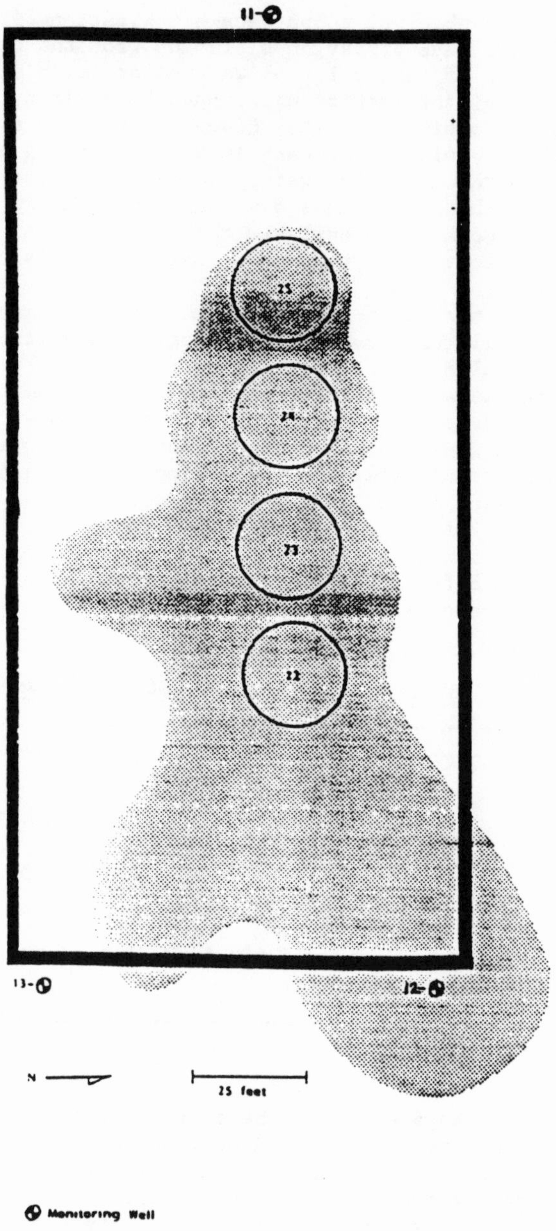

Figure C-1. Total Hydrocarbon Concentrations at a Site with
24-foot Diameter Waste Jet Fuel Tanks.

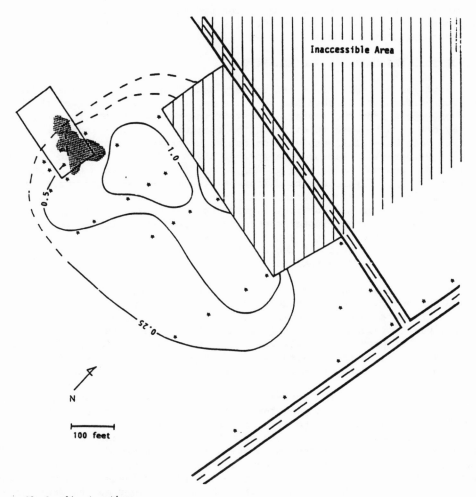

* CO$_2$ Sampling Locations

▓▓ 100 ng/cc Total Petroleum Hydrocarbons

Figure C-2. Results of Soil-Gas Survey Measuring Carbon Dioxide
Concentrations.

gas hydrocarbon concentrations while some distance away false negative results
were obtained from consideration of soil-gas hydrocarbon concentrations as
indicators of underlying contamination. Because of the numerous factors that
can influence these *in situ* microbial processes and our lack of detailed
knowledge of them at any site, it is not possible to guess what may have been
the factors responsible for this difference in behavior. In fact, due to the
probable lack of knowledge of subsurface conditions at any site, prediction of
biodegradation of hydrocarbons or conservative behavior is probably not

possible. Therefore, soil-gas QA/QC for hydrocarbons should incorporate components to deal with this potential situation so that it will not destroy the usefulness of the data.

USE OF SOIL GAS DATA FOR VES PLANNING

Soil-gas concentrations can be very useful in planning a VES. Because optimum performance is obtained when the vapor-withdrawal well of the VES removes vapors from the center of mass of the contamination, the soil-gas survey data can be used to estimate the optimum horizontal location for it. At that point, vertical exploration can be undertaken to optimize the vertical zone of withdrawal. Because the goal of a VES system is to treat soil contamination, and not underlying ground water, it is important to be able to interpret the soil-gas data to differentiate between soil and ground-water contamination as a vapor source. This can be done through consideration of both the magnitudes and horizontal gradients of soil-gas concentrations. Near soil contamination, horizontal concentration gradients will be very steep and concentrations will be high, since they are from pure product and have not been diluted by diffusion of the atmosphere to as large a degree as those resulting from deeper contamination.

At a major aerospace firm in southern California, a vapor degreaser had leaked TCE and PCE into the vadose zone for an unknown period of time. Because of contamination of nearby municipal supply wells, a leak-detection program was implemented and the leak was noted. Due to that, a soil-gas survey at that part of the site was conducted. The site is an industrial facility that has been operated parties since the 1930s. The vadose zone is approximately 120 feet thick and is covered totally by pavement or buildings.

The first part of the survey was a survey to estimate the areal extent of contamination, in order to plan the soil-venting system to be installed there. Figure C-3 shows a map of the site, and Figure C-4 shows the soil-gas PCE concentrations measured at a depth of 4 feet. Several features of Figure C-4 are noteworthy. First, a second source area was indicated near Building 183 and a possible third source area can also be seen. Building 183 is a hazardous-waste storage shed with a gravel-filled drainage basin surrounding it. Surface spills could be hosed from the inside floor out to drain through these basins, since a 4-inch ventilation gap between the bottoms of the building's walls and the floor exists there. A second point that can be noted is that there is considerable horizontal vapor migration evident from these source areas, although they both exhibit the steep horizontal soil-gas concentration gradients characteristic of shallow soil contamination. A third point that can be noted is the seemingly artificial division between the high-concentration zone in the center of Building 175 and the one at the western end. This distinction was made on the basis of soil-gas data for other compounds. Figure C-5 shows the 1,1,1-trichloromethane (TCA) soil-gas data. Because of the much lower concentrations of TCA encountered, the presence of three distinct sources of VOC vapors can be clearly seen. In addition, it can be seen that the TCA/TCE ratios are not the same among the source areas. Based on the results of the above areal delineation of contamination, evaluation of the vertical extent of contamination at locations indicated to be sources was planned. This effort had several objectives: the primary goal was to estimate as closely as possible the vertical center of mass of contamination

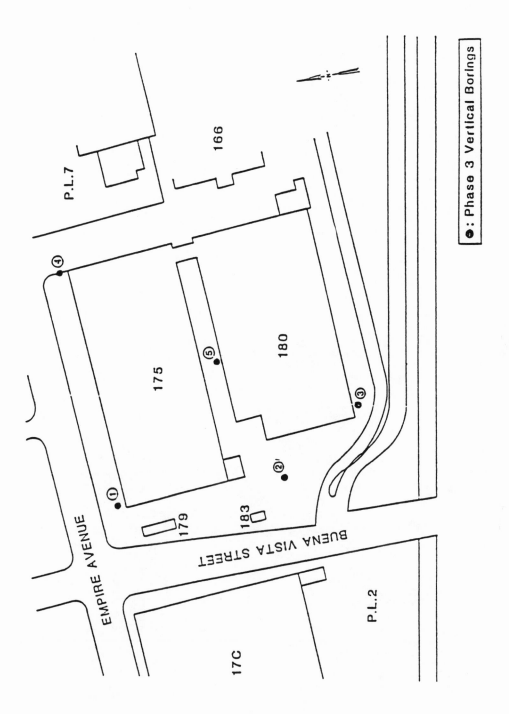

Figure C-3.   Site Map of Industrial Facility.

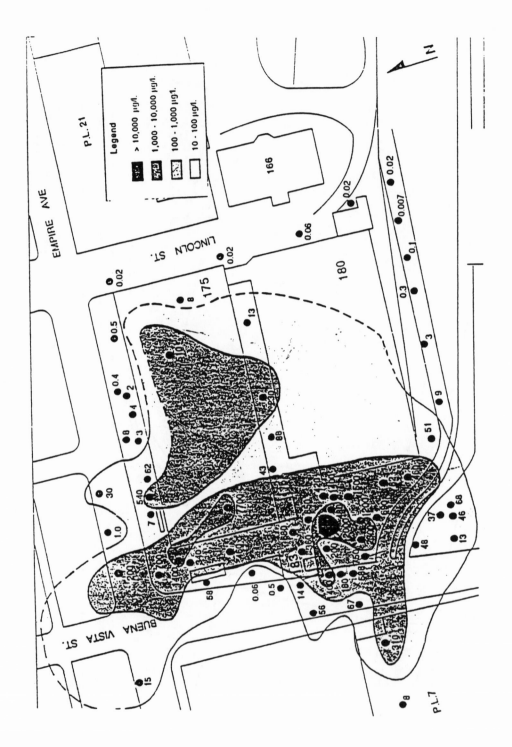

Figure C-4.   PCE Concentrations Measured at a Depth of 4 Feet.

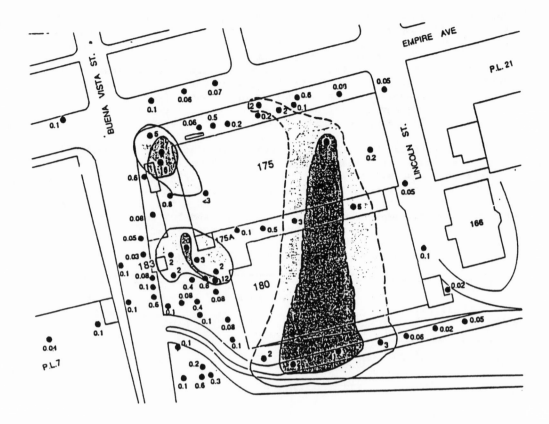

Figure C-5.   TCA Soil-Gas Data.

and to obtain other data valuable for design of the soil-venting system. A
secondary goal was to obtain actual soil samples, in order to determine what
correlation (if any) exists between the soil-gas data and traditional soil-
analysis results. A third objective was to evaluate "background" soil-gas and
soil concentration profiles.

Borings were installed at the five locations shown in Figure C-3.
Borings 1 and 2 were placed to investigate conditions at those two source
areas, while boring 5 was placed to investigate the other anomaly. Due to
limited access to classified areas indoors, boring 5 had to be placed outside
of Building 175. Because of the potential for a source inside Building 180,
boring number 3 was drilled just to the south of it. Building 180 was also a
highly classified facility. Boring number 4 was to evaluate "background" soil
and soil-gas concentrations at the site.

The borings were made with a hollow-stem auger; each 10 feet a soil-gas
probe was attached to A-rod and hammered into the vadose zone 3 feet past the
auger bit. Soil-gas samples were withdrawn and analyzed until duplicates were

within 20 percent of each other. After soil-gas samples were taken and analyzed, soil samples were taken, the soil characterized, and samples wrapped and stored on ice. Samples selected on the basis of the soil-gas data were brought to a laboratory for analysis. At all of the borings, 1/4-inch O.D. tubes attached to a 1-inch perforated metal sampling manifold were installed between two 1 1/2-foot bentonite layers in a 5-foot layer of sand at the bottom of the borehole for future monitoring of subsurface pressure andsoil-gas concentrations and 20 feet above the zone of soil-gas removal. At borings 1, 2, 3, and 5 PVC vapor wells were installed for future soil-venting work.

Table C-4 lists the soil-gas and soil analytical data for boring 1. It can be very clearly seen that the maximum contamination, as indicated by soil-gas concentrations, was at 70 feet. Therefore, 1 1/2-foot bentonite layers were put at depths of 81 and 55 feet with a capped 4-inch PVC pipe terminating at 80 feet, and slotted for 20 feet. As mentioned above, a 1/4-inch O.D. metal tube led to sampling manifolds at 30 feet and 105 feet in 5-foot layers of sand with 1 1/2-foot bentonite layers above and below.

TABLE C-4.   BORING #1 SOIL AND SOIL-GAS DATA.

Depth(ft)	Soil Type (USCS)	Soil-Gas Concentrations (ng/cc)			Soil Concentrations (mg/kg)		
		TCA	TCE	PCE	TCA	TCE	PCE
3	SW	0.6	<0.03	67	<0.01	<0.01	0.013
10	SW	.30	<0.03	15	N/A	N/A	N/A
20	SW	2	<3	380	N/A	N/A	N/A
30	SW	18	<3	790	<0.01	<0.01	0.11
40	SW	40	<0.3	930	N/A	N/A	N/A
50	SW	N/R	N/R	1200	<0.01/ <0.01	<0.01/ <0.01	0.21/ 0.27
60	SW	<3	<3	31000	N/A	N/A	N/A
70	SP	N/R	N/R	108000	N/A	N/A	N/A
80	SP	N/R	N/R	160	<0.01	<0.01	<0.049
90	SP	<0.6	<0.6	370	<0.01	<0.01	0.049
100	SP	2	<0.3	30	<0.01	<0.01	0.012
110	ND	<0.2	<0.3	15	ND	ND	ND

Boring number 2 was drilled to 100 feet and Table C-5 shows the soil-gas and soil data. Boring number 2 was completed similarly to boring number 1 with the placement of the monitoring devices and the 20-foot screened section of PVC pipe based on the soil-gas results.

Boring number 3 was augured to a depth of 70 feet and was completed similarly to borings number 1 and number 2, except that the screened interval of PVC pipe was between 25 and 45 feet. Soil-gas concentrations of TCA and PCE were consistent with remote shallow soil contamination as the source of these compounds. Boring number 5 was drilled to 70 feet also and the slotted interval of PVC pipe was at 20 feet to 40 feet. Table C-6 shows the boring number 3 soil-gas and soil data.

It should be noted that, at boring number 5, soil-gas PCE concentrations of nearly 100 ng/cc were measured at locations with non-detectable soil concentration (below 10 g/kg), while soil-gas PCE concentrations of 5 ng/cc was present at a depth with a soil concentration of 22 g/kg. It is not clear what the reason is for this inconsistent behavior, but it may be related to PCE being partitioned more into the soil-sorbed phase in boring number 3 or to it being trapped in pores where it is not available to contribute to the observed soil-gas concentrations. In addition, the possibility of

TABLE C-5.   BORING #2 SOIL AND SOIL-GAS DATA.

Depth(ft)	Soil Type (USCS)	Soil-Gas Concentrations (ng/cc)			Soil Concentrations (mg/kg)		
		TCA	TCE	PCE	TCA	TCE	PCE
1	SP	0.04	0.3	10	N/A	N/A	N/A
10	SP	<2	<3	2500	<0.01	<0.01	<0.01
20	SP	<4	<8	3900	N/A	N/A	N/A
30	SW/SM	<2	<3	1400	<0.01	<0.01	0.01/ 0.017
40	SW/SM	<2	14	4200	N/A	N/A	N/A
50	GP/SP	<4	33	11000	<0.01	<0.01	<0.01
60	GP/SP	<4	28	5700	N/A	N/A	N/A
72	GP/SP	-	-	--	<0.01	<0.01	<0.01
80	GP/SP	<2	<3	180	N/A	N/A	N/A
90	GP/SP	<0.4	<0.8	3	<0.01	<0.01	<0.01

TABLE C-6.   BORING #3 SOIL AND SOIL-GAS DATA.

Depth(ft)	Soil Type (USCS)	Soil-Gas Concentrations (ng/cc)			Soil Concentrations (mg/kg)		
		TCA	TCE	PCE	TCA	TCE	PCE
2	SP	2	<0.01	1	N/A	N/A	N/A
10	SW	5	<0.01	2	N/A	N/A	N/A
20	SW	7	<0.04	5	0.011	<0.01	0.022
30	SM	2	<0.04	3	N/A	N/A	N/A
40	SM	5	<0.04	5	<0.01	<0.01	<0.01
50	SW	0.04	<0.04	0.4	<0.01	<0.01	<0.01
60	SW	0.2	<0.1	1	N/A	N/A	N/A
70	SW	0.07	<0.04	0.6	<0.01	<0.01	<0.01

experimental error must also be considered. However, from the data it can be seen that this approach allowed for immediate placement of the slotted interval in the zone of highest soil-gas and soil contamination so as to remove contaminant from the center of mass of the contamination. The above data do not describe the complete Initial Remedial Measures that were underway or any interactions between the VES and the ground-water cleanup that was in operation at the site. However, these results do not show the utility of real-time data in planning and installing a Vapor Extraction System.

CONCLUSIONS

Soil-gas sampling and analysis is a potentially rapid and cost-effective approach for delineation of the areal and/or vertical extent of subsurface contamination by volatile organic compounds, and can provide useful data for planning in situ soil stripping systems. Soil-gas data can be used to map plumes, indicate sources, and evaluate subsurface gas-phase dynamics. However, the relationship of soil-gas concentrations to pore-water concentrations, soil-sorbed concentrations, or the presence of residual contaminant are not simple and are probably greatly influenced by subsurface heterogeneity and kinetic factors.

Static soil-gas concentrations can be affected by compound-specific factors, such as biodegradability, as well as site-specific ones, such as barriers to vapor transport. For these reasons, it is important to obtain data describing the relationship between soil-gas and other concentrations on a site-specific basis. In addition, static soil-gas conditions may be quite different than those that will exist during a soil vapor extraction effort.

The typical soil-gas survey for placement of vapor extraction systems involves an areal survey with vertical profiling at selected locations. If this vertical profiling is deep enough, the real-time data obtained can be used to select the depths from which vapors will be withdrawn and hardware for that purpose can be left in the hole. Because of the potential for significant variability in the relationships between soil-gas and soil concentrations within a single site, simplistic use of soil-gas concentrations to estimate the total mass of contaminant present in the vadose zone should be avoided in favor of a carefully planned program to obtain the data needed to make an estimate with a known degree of uncertainty.

Further work is required to more fully understand this technology so that it can be effectively applied. In particular, the mechanism of cleanup (stripping vs. mineralization) for hydrocarbons should be investigated to evaluate the potential for use of the technology for low-volatility hydrocarbons and the effects of temperature should be evaluated. In addition, methods to set cleanup criteria for these types of cleanup should be developed om the basis of the mechanism of a VES cleanup, so that the efficiency of the system as gauged by process-control monitoring data can be used, rather than calculations from unvalidated models. Further progress in the design of VES systems is required to deal with the problem of slow water-phase mass transport of contaminants in the capillary fringe and may be dealt with by pulsed sparging of underlying ground water.

Appendix D
In Situ Biodegradation of Petroleum Distillates in the Vadose Zone

Robert E. Hinchee[a], Douglas C. Downey[b], and Ross N. Miller[c]

Thousands of underground storage tank sites are contaminated with petroleum hydrocarbons. Most of these hydrocarbons are biodegradable if naturally occuring microorganisms are provided an adequate supply of oxygen and basic nutrients. Conventional methods of enhancing natural biodegradation use water to carry oxygen or an alternative electron acceptor to the contamination. However, this method requires substantial amounts of water, and is frequently unsuccessful in unsaturated soil because of poor oxygen distribution.

Using air as the oxygen source is a potentially cost-effective way to increase the microbial degradation of fuel hydrocarbons in unsaturated soil. Biodegradation of hydrocarbons in unsaturated zones through forced aeration has been observed at several field sites. Air must flow through hydrocarbon-contaminated soil at rates and configurations that ensure adequate oxygenation and minimize hydrocarbon-contaminated offgases. Adding nutrients and moisture may be desirable to increase biodegradation rates.

INTRODUCTION

As a result of regulations requiring investigation of underground storage tanks, literally thousands of sites have been identified as contaminated with petroleum hydrocarbons. To date, much attention has been given to pump-and-treat remedial technologies, but this technique leaves a substantial fuel residue in the capillary fringe or vadose zone. Methods of uniformly destroying fuel hydrocarbons in situ, without excessive groundwater pumping or toxic releases to the atmosphere, need to be developed. This paper focuses on one such emerging technology.

Petroleum distillate fuel hydrocarbons are generally biodegradable if naturally occuring microorganisms are provided an adequate supply of oxygen and basic nutrients. Although natural biodegradation does occur, and at many sites will eventually minimize most fuel contamination, the process is frequently too slow to prevent the spread of contamination. Such sites require rapid removal of the contaminant source and groundwater treatment to

[a] Battelle Memorial Institute, 505 King Avenue, Columbus, Ohio 43201
[b] HQ, AFESC/RDVW, Tyndall Air Force Base, Florida 32403
[c] 649 N. 200 E., Kaysville, Utah 84037

protect sensitive aquifers. At these sites, an acceleration or enhancement of the natural biodegradation process is desired.

Important in any in situ remediation is an understanding of the distribution of contaminants. Most of the residue of hydrocarbons at a fuel contaminated site is found in the unsaturated zone soils in the capillary fringe and immediately below the water table. Typically, seasonal water table fluctuations spread residues in the area immediately above and below the water table. Any successful bioremediation effort must treat these areas.

CONVENTIONAL ENHANCED BIODEGRADATION

Over the past two decades the practice of enhanced biodegradation has increased, particularly for treating soluble fuel components in groundwater (Lee et al., 1988). Until recently, less emphasis was given to enhancing biodegradation in the unsaturated zone. The current conventional enhanced bioreclamation process uses water to carry the oxygen or an alternative electron acceptor to the contamination, whether it occurs in the groundwater or unsaturated zone.

A recent field experiment at a jet fuel-contaminated site using infiltration galleries and spray irrigation to introduce oxygen, nitrogen and phosphorous to unsaturated, sandy soils was unsuccessful due to rapid H_2O_2 decomposition and resulting poor oxygen distribution (Hinchee et al., 1989a). A study being conducted by the U.S. Environmental Protection Agency and the U.S. Coast Guard at Traverse City, Michigan, uses deep well injection to raise the water table in order to supply oxygen-enriched water to the contaminated soils. Pure oxygen and hydrogen peroxide have been used as oxygen sources, and recently nitrate has been added as an alternative to oxygen. Preliminary results indicate better hydrogen peroxide stability than achieved by Hinchee et al. (1989a). Some degradation of the aromatic hydrocarbon appears to have occured; however, total hydrocarbon contamination levels appear unaffected (Ward, 1988).

In most cases where water is used as the oxygen carrier, oxygen is the limiting factor for biodegradation. If pure oxygen is utilized and 40 mg/l of dissolved oxygen is achieved, approximately 25,000 lbs of water must be delivered to the formation to degrade a single pound of hydrocarbon. If 500 mg/l of hydrogen peroxide is successfully delivered, then 4,000 lbs of water are necessary. As a result, even if hydrogen peroxide can be successfully used, substantial volumes of water must be pumped through the contaminated formation to deliver sufficient oxygen.

ENHANCED BIODEGRADATION THROUGH SOIL VENTING

A system engineered to increase the microbial degradation fo fuel hydrocarbons in the vadose zone using forced air as the oxygen source is a potentially cost-effective alternative to conventional systems. This process stimulates soil-indigenous microorganisms to aerobically metabolize fuel hydrocarbons in unsaturated soils. Depending upon air flow rates, volatile compounds may be simultaneously removed from contaminated soils.

By using air as an oxygen source, the minimum ratio of air pumped to hydrocarbon degraded is approximately 13 lb to 1 lb. This compares to well over 1000 lb of water per lb of hydrocarbon for a conventional waterborne enhanced bioreclamation process. An additional advantage of using an airborne process is that gases have greater diffusivity than liquids. At many sites, geological heterogeneities present an added problem with a waterborne oxygen source, because fluid pumped through the formation is channeled into the more permeable pathways. For example, in an alluvial soil with interbedded sand and clay, virtually all of the fluid flow will take place in the sand. As a result, oxygen must be delivered to the less permeable clay lenses through diffusion. In a gaseous system, this diffusion can be expected to take place at a rate several orders of magnitude greater than in a liquid system. While it is probably not realistic to expect diffusion to aid in water-based bioreclamation, in an air-based application, diffusion may be a significant mechanism for oxygen delivery.

The first documented evidence of unsaturated zone biodegradation resulting from forced aeration was reported by the Texas Research Institute, Inc., in a study for the American Petroleum Institute. A large-scale model experiment was conducted to test the effectiveness of a surfactant treatment to enhance recovery of spilled gasoline. The experiment accounted for only 8 of the 65 gallons originally spilled and raised questions about the fate of the gasoline. A subsequent column study was conducted to determine a diffusion coefficient for soil venting. This column study evolved into a biodegradation study in which it was concluded that as much as 38% of the fuel hydrocarbon was biologically mineralized. Researchers concluded that venting would not only remove gasoline by physical means, but also could enhance microbial activity (Texas Research Institute, 1980; Texas Research Institute, 1984).

Wilson and Ward (1986) suggest that using air as a carrier for oxygen could be 1000 times more efficient than transferring it to water, especially in deep, hard-to-flood unsaturated zones. They made the connection between soil venting and biodegradation by observing that, "soil venting uses the same principle to remove volatile components of the hydrocarbon." In a general overview of the soil venting process, Bennedsen et al. (1987) concluded that soil venting provides large quantities of oxygen to the vadose zone, possibly stimulating aerobic degradation. He states that water and nutrients would also be required for significant degradation and encourages additional investigation into this area.

Biodegradation enhanced by soil venting has been observed at several field sites. Investigators at a soil venting site for remediation of gasoline contaminated soil claim significant biodegradation as measured by a temperature rise that they attributed to biodegradation. They claim that the pile was cleaned up during the summer primarily by biodegradation (Conner, 1988). However, they did not control for natural volatilization from the above ground pile, and not enough data was provided to critically review the biodegradation claim.

Researchers at Traverse City, Michigan, measured toluene concentration in vadose zone soil gas as an indicator of fuel contamination in the vadose zone. They assumed absence of advection and attributed toluene loss to

biodegradation. The investigators conclude that, because toluene
concentrations decayed near the oxygenated ground surface, soil venting is an
attractive remediation alternative for biodegrading light volatile hydrocarbon
spills (Ostendorf and Kambell, 1989).

Hinchee et al. (1989b) documented biodegradation of JP-4 at a
conventional soil venting site. Biodegradation accounted for at least 15% of
the hydrocarbon removal. Their study compared O_2 and CO_2 invented gases from
the contaminated area to that from an uncontaminated area. Additionally, they
measured the stable carbon isotope makeup of the CO_2 carbon to confirm the JP-
4 as its source.

Ely and Heffner (1988) of the Chevron Research Company patented a
process for the in situ biodegradation of spilled hydrocarbons using soil
venting. Experimental design and data are not provided, but findings are
presented graphically. At a gasoline and diesel oil contaminated site,
results indicated that about 2/3 of the hydrocarbon removal was by
volatilization and 1/3 by biodegradation. At a site containing only fuel
oils, approximately 20 gal/well/day were biodegraded, while vapor pressures
were too low for removal by volatilization. Ely and Heffner (1988) claim that
the process is more advantageous than strict soil venting because removal is
not dependent only on vapor pressure. In the examples stated in the patent,
CO_2 was maintained between 6.8% and 11% and O_2 between 2.3% and 11% in vented
air. The patent suggests that the addition of water and nutrients may not be
acceptable because of flushing to the water table, but nutrient addition is
claimed as part of the patent. The patent recommends flow rates between 30
and 250 cfm per well and states that air flows higher than those required for
volatilization may be optimum for biodegradation.

APPLICATIONS

The use of an air-based oxygen supply for enhancing biodegradation
relies upon air flow through hydrocarbon contaminated soils at rates and
configurations that will ensure adequate oxygenation of aerobic biodegradation
and minimize or eliminate the production of a hydrocarbon contaminated offgas.
The addition of nutrients and moisture may be desirable to increase
biodegradation rates. Figures 1 and 2 illustrate possible air
injection/withdrawal configurations. Dewatering is illustrated in the
figures, but this may not always be necessary depending upon the distribution
of contaminants relative to the water table. However, it is required at many
fuel hydrocarbon contaminated sites. A key feature not illustrated is the
narrowly screened soil gas monitoring points that sample only a short vertical
section of the soil. These points are required to determine local oxygen
concentrations. Measurements of oxygen levels in the vent are not
representative of local conditions. Nutrient and moisture addiction typically
may take any of a variety of configurations.

The configuration in Figure D-1 is more like a conventional soil venting
installation where air is drawn from a vent well in the area of greatest
contamination. The advantage of this configuration is that it generally
requires the least air pumping; the disadvantage is that hydrocarbon offgas
concentration is probably maximized and all of the capillary fringe
contamination may not be treated. Figure D-2 illustrates a configuration in

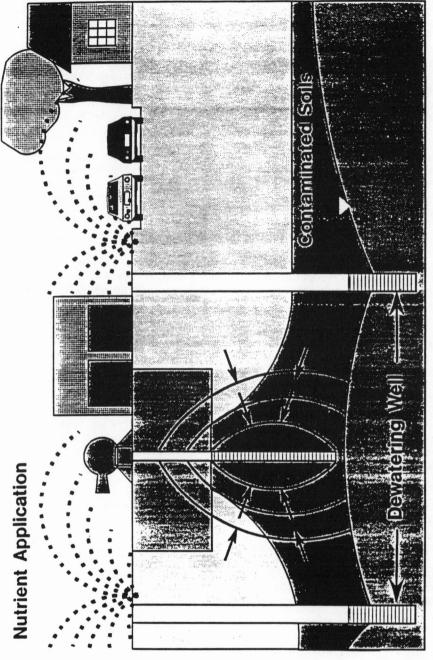

Figure 1. Potential Configuration for Enhanced Bioreclamation Through Soil Venting (Air Withdrawn From Vent Well)

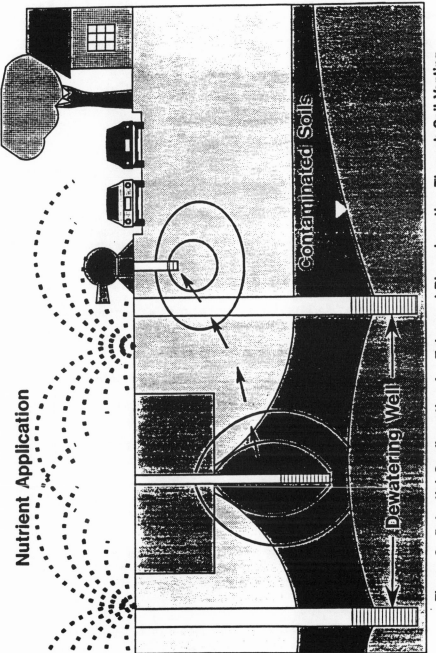

Figure 2. Potential Configuration for Enhanced Bioreclamation Through Soil Venting (Air Withdrawn From Clean Soil)

which air is injected into the contaminated zoneand withdrawn from clean soils. This configuration allows the more volatile hydrocarbons to degrade prior to being withdrawn and thereby eliminates contaminated offgases. The optimal configuration for any given site will, of course, depend upon site-specific conditions and remedial objectives.

The significant features of this technology include:

- Optimizing air flow to reduce volatilization while maintaining aerobic conditions for biodegradation.

- Monitoring local soil gas conditions to assure aerobic conditions, not just monitoring vent gas composition.

- Adding moisture and nutrients as required for air/contaminant contact.

The U.S. Air Force Engineering and Service Laboratory presently is sponsoring a field-scale pilot test of the enhanced biodegradation technology. The site is JP-4 contaminated and air is being introduced and withdrawn in configurations similar to those shown in Figures 1 and 2. The effect of varying nutrients and soil moisture levels is being evaluated. The use of conventional soil venting to enhance biodegradation is being practiced on a limited basis at a few other sites; however, at most soil venting sites, the biodegradation effects are ignored and undocumented.

RECOMMENDATIONS

Soil venting alone, with no nutrient or moisture addition, typically results in some stimulation of in situ biodegradation. The following recommendations are made for documenting biodegradation when conducting conventional soil venting of fuel hydrocarbon contaminated soils:

1. Prior to venting, determine soil gas hydrocarbon, CO_2, and O_2 profiles.

2. Determine the end point for venting. A mixed hydrocarbon fuel such as JP-4 has a fraction too heavy to volatilize, and it is possible that biodegradation may continue after the light end has volatilized.

3. Develop an estimate of noncontaminant respiration. This may be accomplished either through background measurements of CO_2 and O_2 in an uncontaminated location or by means of carbon isotopic analysis.

To further pursue the development of an enhanced in situ soil-venting technology, the following studies are recommended:

1. Fuel biodegradation in unsaturated soils to develop a better understanding of variables such as oxygen content, nutrient requirements, soil moisture, contaminant levels (both high end for

       possible toxic effects and low end for treatment limits), and soil
       types

2.   Gas transport in the vadose zone to allow adequate design of air
       delivery system

3.   Nutrient and moisture delivery systems, including possible gaseous
       nutrient injection (i.e., NH_3); means of engineering moisture
       addition in deeper stratified formations; and nutrient
       formulations to allow adequate nutrient mobility in pore water

REFERENCES

Bennedsen, M.B., Scott, J.P., and Hartley, J.D. 1987. Use of Vapor Extraction Systems for In Situ Removal of Volatile Organic Compounds from Soil. Proceedings of National Conference on Hazardous Wastes and Hazardous Materials, Washington, D.C. pp. 92-95.

Conner, R.J. 1988. Case Study of Soil Venting. Poll. Eng. 7:74-78.

Ely, D.L. and Heffner, D.A. 1988. Process for In Situ Biodegradation of Hydrocarbon Contaminated Soil. U.S. Patent Number 4,765,902.

Hinchee, R.E., Downey, D.C., Dupont, R.R., and Arthur, M.F. 1989a. Enhanced Biodegradation through Soil Venting. Unpublished Report Submitted to the U.S. Air Force Engineering and Services Laboratory.

Hinchee, R.E., Downey, D.C., Slaughter, J.K., and Westray, M. 1989b.Enhanced Biorestoration of Jet Fuels; A Full Scale Test at Elgin Air Force Base, Florida. Air Force Engineering and Services Center Report ESL/TR/88-78.

Lee, M.D., et al. 1988. Biorestoration of Aquifers Contaminated with Organic Compounds. CRC Critical Reviews in Env. Control, Vol. 18 pp. 29-89.

Ostendorf, D.W. and Kambell, D.H. 1989. Vertical Profiles and Near Surface Traps for Field Measurement of Volatile Pollution in the Subsurface Environment. Proceedings of NWWA Conference on New Techniques for Quantifying the Physical and Chemical Properties of Heterogeneous Aquifers, Dallas, TX.

Texas Research Institute. 1980. Laboratory Scale Gasoline Spill and Venting Experiment. American Petroleum Institute, Interim Report No. 7743-5:JST.

Texas Research Institute. 1984. Forced Venting to Remove Gasoline Vapor from a Large-scale Model Aquifer. American Petroleum Institute. Final Report No. 82101-F:TAV.

Ward, C.H. 1988. A Quantitative Demonstration of the Raymond Process for In Situ Biorestoration of Contaminated Aquifers. Proceedings of NWWA/API Conference on Petroleum Hydrocarbons and Organic Chemicals in Groundwater, Houston, TX pp 723-746.

Wilson, J.T., and Ward, C.H. 1986. Opportunities for Bioremediation of Aquifers Contaminated with Petroleum Hydrocarbons. J. Ind. Microbiol., 27:109-116.

Appendix E
A Practical Approach to the Design, Operation, and Monitoring of In-Situ Soil Venting Systems

P. C. Johnson[a], M. W. Kemblowski[a], J. D. Colthart[a],
D. L. Byers[a], and C. C. Stanley[b]

INTRODUCTION

When operated properly, in-situ soil venting or vapor extraction can be one of the more cost-effective remediation processes for soils contaminated with gasoline, solvents, or other relatively volatile compounds. A "basic" system, such as that shown in Figure E-1, couples vapor extraction (recovery) wells with blowers or vacuum pumps to remove vapors from the vadose zone and thereby reduce residual levels of soil contaminants. More complex systems incorporate trenches, air injection wells, passive wells, and surface seals. Above-ground treatment systems condense, adsorb, or incinerate vapors; in some cases vapors are simply emitted to the atmosphere through diffuser stacks. In-situ soil venting is an especially attractive treatment option because the soil is treated in place, sophisticated equipment is not required, and the cost is typically lower than other options.

The basic phenomena governing the performance of soil venting systems are easily understood. By applying a vacuum and removing vapors from extraction wells, vapor flow through the unsaturated soil zone is induced. Contaminants volatilize from the soil matrix and are swept by the carrier gas flow (primarily air) to the extraction wells or trenches. Many complex processes occur on the microscale, however, the three main factors that control the performance of a venting operation are the chemical composition of the contaminant, vapor flowrates through the unsaturated zone, and the flowpath of carrier vapors relative to the location of the contaminants.

The components of soil venting systems are typically off-the-shelf items, and the installation of wells and trenches can be done by most reputable environmental firms. However, the design, operation, and monitoring of soil venting systems is not trivial. In fact, choosing whether or not venting should be applied at a given site is a difficult question in itself. If one decides to utilize venting, design questions involving the number of

[a]Shell Development/[b]Shell Oil Company, Westhollow Research Center
 Houston, TX 77152-1380

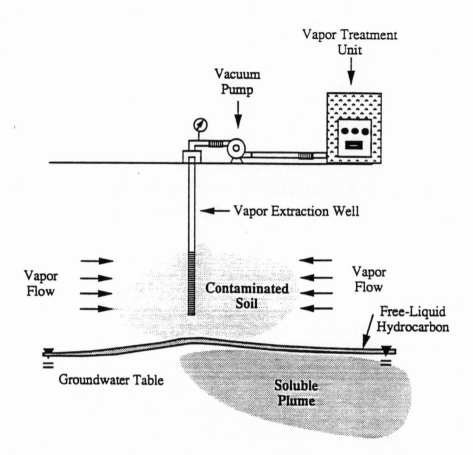

Figure E-1.   "Basic" In-Situ Soil Venting System

wells, well spacing, well location, well construction, and vapor treatment systems must then be answered. It is the current state-of-the-art that such questions are answered more by instinct than by rigorous logic. This is evidenced by the published soil venting "success stories" (see Hutzler et al.[1] for a good review), which rarely include insight into the design process.

In this paper we suggest a series of steps and questions that must be followed and answered in order to a) decide if venting is appropriate at a given site, and b) to design cost-effective in-situ soil venting systems. This series of steps and questions forms a "decision tree" process that could be easily incorporated in a PC-based expert system. In the development of this approach we will attempt to identify the limitations of in-situ soil venting, and subjects or behavior that are difficult to quantify and for which future study is needed.

THE "PRACTICAL APPROACH"

Figure E-2 presents a flowchart of the process discussed in this paper. Each step of the flowchart is discussed below in detail, and where appropriate, examples are given.

The Site Investigation

Whenever a soil contamination problem is detected or suspected, a site investigation is conducted to characterize and delineate the zone of soil and groundwater contamination. Often the sequence of steps after initial response and abatement is as follows:

(a) *background review*: Involves assembling historical records, plot plans, engineering drawings (showing utility lines), and interviewing site personnel. This information is used to help identify the contaminant, probable source of release, zone of contamination, and potentially impacted areas (neighbors, drinking water supplies, etc.).

(b) *preliminary site screening*: Preliminary screening tools such as soil-gas surveys and cone penetrometers are used to roughly define the zone of contamination and the site geology. Knowledge of site geology is essential to determine probable migration of contaminants through the unsaturated zone.

(c) *detailed site characterization*: Soil borings are drilled and monitoring wells are installed based on the results from steps (a) and (b).

(d) *contaminant characterization*: soil and groundwater samples are analyzed to determine contaminant concentrations and compositions.

Costs associated with site investigations can be relatively high depending on the complexity of the site and size of the spill or leak. For large spills and complex site geological/hydrogeological conditions, site investigation costs are often comparable to remediation costs. In addition, the choice and design of a remediation system is based on the data obtained during the site investigation. For these reasons it is important to insure that specific information is collected, and to validate the quality of the data.

If it is presumed that in-situ soil venting will be a candidate for treatment, then the following information needs to be obtained during the preliminary site investigation:

(a) *site geology* - this includes soil type and subsurface stratigraphy. While they are not essential, the moisture content, total organic carbon, and permeability of each distinct soil layer also provides useful information that can be used to choose and design a remediation system.

(b) *site hydrogeology* - the water table depth and gradient must be known, as well as estimates of the aquifer permeability.

(c) *contaminant composition, distribution and residual levels* - soil samples should be analyzed to determine which contaminants are present at what

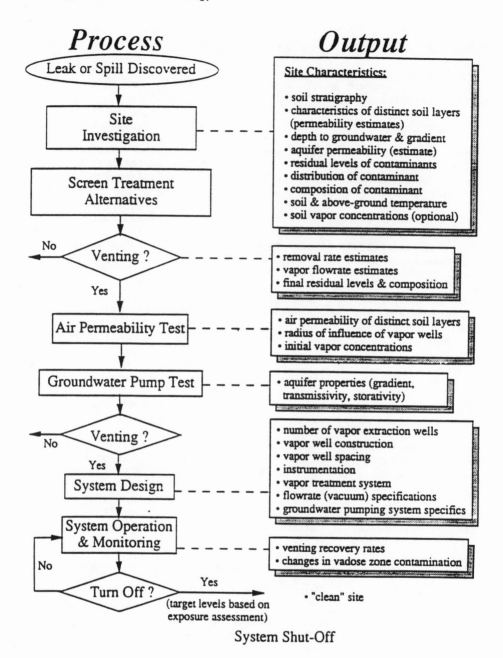

Figure E-2.   In-Situ Soil Venting System Design Process.

levels. Recommended analytical methods should be used to identify target compounds (i.e., benzene, toluene, or xylenes) and total hydrocarbons present. For soil analyses these methods are:

> EPA 8240 - volatile organic chemicals
> EPA 8270 - semi-volatile organic chemicals
> EPA 418.1 - total petroleum hydrocarbons

The corresponding water analyses methods are:

> EPA 624 - volatile organic chemicals
> EPA 625 - semi-volatile organic chemicals
> EPA 418.1 - total petroleum hydrocarbons

With the current high cost of chemical analyses it is important to intelligently select which analyses should be performed and which samples should be sent to a certified laboratory. Local regulations usually require that a minimum number of soil borings be performed, and target compounds must be analyzed for based on the suspected composition of the contamination. Costs can be minimized and more data obtained by utilizing field screening tools, such as hand-held vapor meters or portable field GC's. These instruments can be used to measure both residual soil contamination levels and headspace vapors above contaminated soils. At a minimum, soil samples corresponding to lithology changes or obvious changes in residual levels (based on visual observations or odor) should be analyzed.

For complex contamination mixtures, such as gasoline, diesel fuel, and solvent mixtures, it is not practical or necessary to identify and quantify each compound present. In such cases it is recommended that a "boiling point" distribution be measured for a representative sample of the residual contamination. Boiling point distribution curves, such as shown in Figure E-3 for "fresh" and "weathered" gasoline samples, can be constructed from GC analyses of the soil residual contamination (or free-product) and knowledge of the GC elution behavior of a known series of compounds (such as straight-chain alkanes). Compounds generally elute from a GC packed column in the order of increasing boiling point, so a boiling point distribution curve is constructed by grouping all unknowns that elute between two known peaks (i.e. between n-hexane and n-heptane). Then they are assigned an average boiling point, molecular weight, and vapor pressure. Use of this data will be explained below.

(d) *temperature* - both above- and below-ground surface.

The cone penetrometer, which is essentially an instrumented steel rod that is driven into the soil, is becoming a popular tool for preliminary site screening investigations. By measuring the shear and normal forces on the leading end of the rod, soil structure, and hence permeability can be defined. Some cone penetrometers are also constructed to allow the collection of vapor or groundwater samples. This tool has several advantages over conventional soil boring techniques (as a preliminary site characterization tool): the subsurface soil structure can be defined better, no soil cuttings are generated, and more analyses can be performed per day.

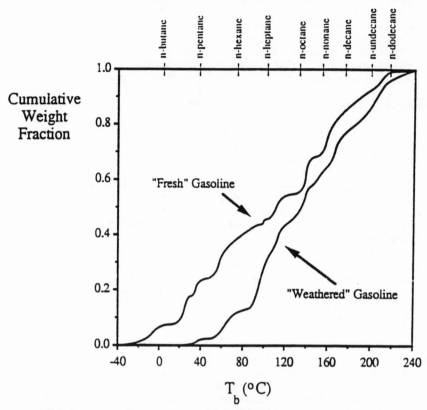

Figure E-3.   Boiling Point Distribution Curves for Samples of
"Fresh" and "Weathered" Gasolines.

Results from the preliminary site investigation should be summarized in contour plots, fence diagrams, and tables prior to analyses.

<u>Deciding if Venting is Appropriate</u>

As stated above, the three main factors governing the behavior of any in-situ soil venting operation are the vapor flow rate, contaminant vapor concentrations, and the vapor flowpath relative to the contaminant location. In an article by Johnson et al.[2] simple mathematical equations were presented to help quantify each of these factors. Below we illustrate how to utilize these "screening models" and the information collected during the preliminary site investigation to help determine if in-situ soil venting is appropriate at a given site. In making this decision we will answer the following questions:

(1) *What contaminant vapor concentrations are likely to be obtained?*

(2) *Under ideal vapor flow conditions (i.e. 100 - 1000 scfm vapor
   flowrates), is this concentration great enough to yield acceptable
   removal rates?*

(3) What range of vapor flowrates can realistically be achieved?

(4) Will the contaminant concentrations and realistic vapor flowrates produce acceptable removal rates?

(5) What are the vapor composition and concentration changes? What residual, if any, will be left in the soil?

(6) Are there likely to be any negative effects of soil venting?

Negative answers to questions (2), (3), or (4) will rule out in-situ soil venting as a practical treatment method.

(1) - What contaminant vapor concentrations are likely to be obtained?

Question (1) can be answered based on the results of soil vapor surveys, analyses of headspace vapors above contaminated soil samples, or equilibrium vapor models[2]. In some cases just knowing which compounds are present is sufficient to estimate if venting is feasible. In the absence of soil-vapor survey data, contaminant vapor concentrations can be estimated. The maximum vapor concentration of any compound (mixture) in extracted vapors is its equilibrium or "saturated" vapor concentration, which is easily calculated from knowledge of the compound's (mixture's) molecular weight, vapor pressure at the soil temperature, residual soil contaminant composition, and the ideal gas law:

$$C_{est} = \sum_i \frac{x_i P_i^v M_{v,i}}{RT}$$

(E-1)

where:

C_{est} — estimate of contaminant vapor concentration [mg/l]

x_i — mole fraction of component i in liquid-phase residual ($x_i = 1$ for single compound)

P_i^v — pure component vapor pressure at temperature T [atm]

$M_{v,i}$ — molecular weight of component i [mg/mole]

R — gas constant = 0.0821 1-atm/mole-°K

T — absolute temperature of residual [°K]

Table E-1 presents data for some chemicals and mixtures often spilled in the environment. There are more sophisticated equations for predicting vapor concentrations in soil systems based on equilibrium partitioning arguments, but these require more detailed information (organic carbon content, soil moisture) than is normally available. If a site is chosen for remediation, the residual total hydrocarbons in soil typically exceed 500 mg/kg. In this residual concentration range the majority of hydrocarbons will be present as a separate or "free" phase, the contaminant vapor concentrations become independent of residual concentration (but still depend on composition), and Equation E-1 is applicable[2]. In any case, it should be noted that these are estimates only for vapor concentrations at the start of venting, which is when the removal rates are generally greatest. Contaminant concentrations in the extracted vapors will decline with time due to changes in composition, residual levels, or increased diffusional resistances. These topics are

Table E-1. Selected Compounds and Their Chemical Properties.

Compound	M_w (g/mole)	T_b (1 atm) (°C)	P_v^o (20°C) (atm)	C_{sat} (mg/l)
n-pentane	72.2	36	0.57	1700
n-hexane	86.2	69	0.16	560
trichloroethane	133.4	75	0.132	720
benzene	78.1	80	0.10	320
cyclohexane	84.2	81	0.10	340
trichloroethylene	131.5	87	0.026	140
n-heptane	100.2	98	0.046	190
toluene	92.1	111	0.029	110
tetrachloroethylene	166	121	0.018	130
n-octane	114.2	126	0.014	65
chlorobenzene	113	132	0.012	55
p-xylene	106.2	138	0.0086	37
ethylbenzene	106.2	138	0.0092	40
m-xylene	106.2	139	0.0080	35
o-xylene	106.2	144	0.0066	29
styrene	104.1	145	0.0066	28
n-nonane	128.3	151	0.0042	22.0
n-propylbenzene	120.2	159	0.0033	16
1,2,4 trimethylbenzene	120.2	169	0.0019	9.3
n-decane	142.3	173	0.0013	7.6
DBCP	263	196`	0.0011	11
n-undecane	156.3	196	0.0006	3.8
n-dodecane	170.3	216	0.00015	1.1
napthalene	128.2	218	0.00014	0.73
tetraethyllead	323	dec. @200C	0.0002	2.6
gasoline[1]	95	-	0.34	1300
weathered gasoline[2]	111	-	0.049	220

[1] Corresponds to "fresh" gasoline defined in Table E-2 with boiling point distribution shown in Figure E-3.

[2] Corresponds to "weathered" gasoline defined in Table E-2 with boiling point distribution shown in Figure E-3.

discussed below in more detail.

(2) - Under ideal vapor flow conditions (i.e. 100 - 1000 scfm vapor
flowrates), is this concentration great enough to yield acceptable
removal rates?

Question (2) is answered by multiplying the concentration estimate C_{est},
by a range of reasonable flowrates, Q:

$$R_{est} = C_{est} \, Q \qquad\qquad (E-2)$$

Here R_{est} denotes the estimated removal rate, and C_{est} and Q must be
expressed in consistent units. For reference, documented venting operations
at service station sites typically report vapor flowrates in the 10 - 100 scfm
range[1], although 100 - 1000 scfm flowrates are achievable for very sandy soils
or large numbers of extraction wells. At this point in the decision process
we are still neglecting that vapor concentrations decrease during venting due
to compositional changes and mass transfer resistances. Figure E-4 presents
calculated removal rates R_{est} [kg/d] for a range of C_{est} and Q values. C_{est}
values are presented in [mg/l] and [ppm$_{CH4}$] units, where [ppm$_{CH4}$] represents
methane-equivalent parts-per-million volume/volume (ppm$_v$) units. The [ppm$_{CH4}$]
units are used because field analytical tools that report [ppm$_v$] values are
often calibrated with methane. The [mg/l] and [ppm$_{CH4}$] units are related by:

$$[mg/l] = \frac{[ppm_{CH4}] * 16000mg\text{-}CH_4/mole\text{-}CH_4 * 10^{-6}}{(0.0821 \ 1\text{-}atm/^{\circ}K\text{-}mole) * (298K)} \qquad (E-3)$$

For field instruments calibrated with other compounds (i.e., butane,
propane) [ppm$_v$] values are converted to [mg/l] by replacing the molecular
weight of CH_4 in Equation E-3 by the molecular weight [mg/mole] of the
calibration compound.

Acceptable or desirable removal rates $R_{acceptable}$, can be determined by
dividing the estimated spill mass M_{spill}, by the maximum acceptable clean-up
time τ:

$$R_{acceptable} = M_{spill}/\tau \qquad\qquad (E-4)$$

For example, if 1500 kg ($\approx$500 gal) of gasoline had been spilled at a
service station and we wished to complete the clean-up within eight months,
then $R_{acceptable} = 6.3$ kg/d. Based on Figure E-4, therefore, C_{est} would have to
average >1.5 mg/l (2400 ppm$_{CH4}$) for Q=2800 l/min (100 cfm) if venting is to be
an acceptable option. Generally, removal rates <1 kg/d will be unacceptable
for most spills, so soils contaminated with compounds (mixtures) having
saturated vapor concentrations less than 0.3 mg/l (450 ppm$_{CH4}$) will not be
good candidates for venting, unless vapor flowrates exceed 100 scfm. Judging
from the compounds listed in Table E-1, this corresponds to compounds with
boiling points (T_b)>150°C, or pure component vapor pressures <0.0001 atm

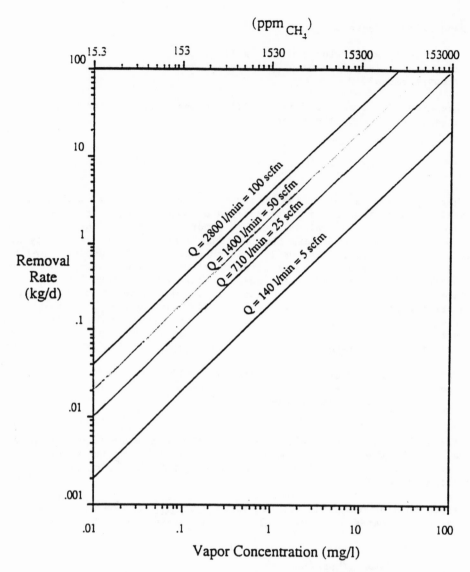

(ppm_{CH_4}) - concentration in methane-equivalent ppm (vol./vol.) units

Figure E-4.   In-Situ Soil Venting Removal Rate Dependence on
Vapor Extraction Rate and Vapor Concentration.

evaluated at the subsurface temperature.

(3) - What range of vapor flowrates can realistically be achieved?

Question (3) requires that we estimate realistic vapor flowrates for our site specific conditions. Equation E-5, which predicts the flowrate per unit thickness of well screen Q/H [cm^3/s], can be used for this purpose:

$$\frac{Q}{H} = \pi \frac{k}{\mu} P_w \frac{[1 - (P_{atm}/P_w)^2]}{\ln(R_w/R_I)} \qquad (E\text{-}5)$$

where:

k	=	soil permeability to air flow [cm^2] or [darcy]
μ	=	viscosity of air = 1.8 x 10^{-4} g/cm-s or 0.018 cp
P_w	=	absolute pressure at extraction well [g/cm-s^2] or [atm]
P_{Atm}	=	absolute ambient pressure $\approx$ 1.01 x 10^6 g/cm-s^2 or 1 atm
R_w	=	radius of vapor extraction well [cm]
R_I	=	radius of influence of vapor extraction well [cm]

This equation is derived from the simplistic steady-state radial flow solution for compressible flow[2], but should provide reasonable estimates for vapor flow rates. If we can measure or estimate k, then the only unknown parameter is the empirical "radius of influence" R_I. Values ranging from 9 m (30 ft) to 30 m (100 ft) are reported in the literature for a variety of soil conditions, but fortunately Equation E-5 is not very sensitive to large changes in R_I. For estimation purposes, therefore, a value of R_I=12 m (40 ft) can be used without a significant loss of accuracy. Typical vacuum well pressures range from 0.95 - 0.90 atm (20 - 40 in H$_2$O vacuum). Figure E-5 presents predicted flowrates per unit well screen depth Q/H, expressed in "standard" volumetric units Q*/H (= Q/H(P_w/P_{Atm})) for a 5.1 cm radius (4" diameter) extraction well, and a wide range of soil permeabilities and applied vacuums. Here H denotes the thickness of the screened interval, which is often chosen to be equal to the thickness of the zone of soil contamination (this minimizes removing and treating any excess "clean" air). For other conditions the Q*/H values in Figure E-5 can be multiplied by the following factors:

R_w = 5.1 cm (2") R_I = 7.6 m (25')   - multiply by Q*/H by 1.09
R_w = 5.1 cm (2") R_I = 23 m (75')   - multiply by Q*/H by 0.90
R_w = 7.6 cm (3") R_I = 12 m (40')   - multiply by Q*/H by 1.08
R_w = 10 cm (4") R_I = 12 m (40')   - multiply by Q*/H by 1.15
R_w = 10 cm (4") R_I = 7.6 m (25')   - multiply by Q*/H by 1.27

As indicated by the multipliers given above, changing the radius of influence from 12 m (40 ft) to 23 m (75 ft) only decreases the predicted flowrate by 10%. The largest uncertainty in flowrate calculations will be due to the air permeability value k, which can vary by one to three orders of magnitude across a site and can realistically only be estimated from boring log data within an order of magnitude. It is prudent, therefore, to choose a range of k values during this phase of the decision process. For example, if boring logs indicate fine sandy soils are present, then flowrates should be

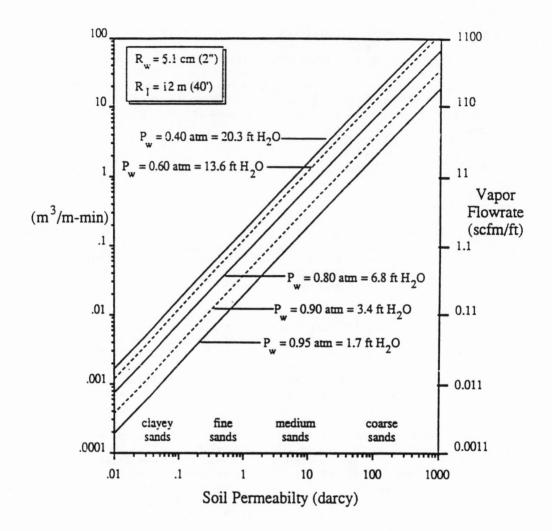

[ft H$_2$O] denote vacuums expressed as equivalent water column heights

Figure E-5.   Predicted Steady-State Flowrates (per unit well screen depth)
for a Range of Soil Permeabilities and Applied Vacuums (P$_v$).

calculated for k values in the range 0.1<k<1.0 darcy.

(4) - Will the contaminant concentrations and realistic vapor flowrates produce acceptable removal rates?

Again, estimated removal rates R_{est}, must be compared with an acceptable rate $R_{acceptable}$, as determined from Equation E-4. Maximum removal rates are achieved when the induced vapor flow travels only through the zone of soil contamination and no mass-transfer limitations are encountered. In other words, all vapor flows through contaminated soils and becomes saturated with contaminant vapors. For this "best" case the estimated removal rate is given by Equation E-2:

$$R_{est} = C_{est}\ Q \qquad\qquad (E\text{-}2)$$

We are still neglecting changes in C_{est} with time due to composition changes. Other less optimal conditions are often encountered in practice and it is useful to be able to quantify how much lower the removal rate will be from the value predicted by Equation E-2. We will consider the three cases illustrated in Figure E-6a, b, and c.

In Figure E-6a, a fraction ϕ of the vapor flows through uncontaminated soil. The fraction can be roughly estimated by assessing the location of the well relative the contaminant distribution. In Figure E-6a, for example, it appears that roughly 25% of the vapor flows through uncontaminated soil. The maximum removal rate for this case is then:

$$R_{est} = (1-\phi)Q\ C_{est} \qquad\qquad (E\text{-}6)$$

In Figure E-6b, vapor flows parallel to, but not through, the zone of contamination, and the significant mass transfer resistance is vapor phase diffusion. This would be the case for a layer of liquid hydrocarbon resting on top of an impermeable strata or the water table. This problem was studied by Johnson et al.[2] for the case of a single component. Their solution is:

$$R_{est} = \eta Q C_{est}$$

$$\eta = \frac{1}{3H}(6D\mu/k)^{1/2}[\ln(R_I/R_v)/(P_{Atm}-P_v)]^{1/2}\ [R_2^2-R_1^2]^{1/2} \qquad (E\text{-}7)$$

where:

η = efficiency relative to maximum removal rate
D = effective soil vapor diffusion coefficient [cm^2/s]
μ = viscosity of air = 1.8×10^{-4} g/cm-s
k = soil permeability to vapor flow [cm^2]
H = thickness of screened interval [cm]
R_I = radius of influence of venting well [cm]
R_v = venting well radius [cm]

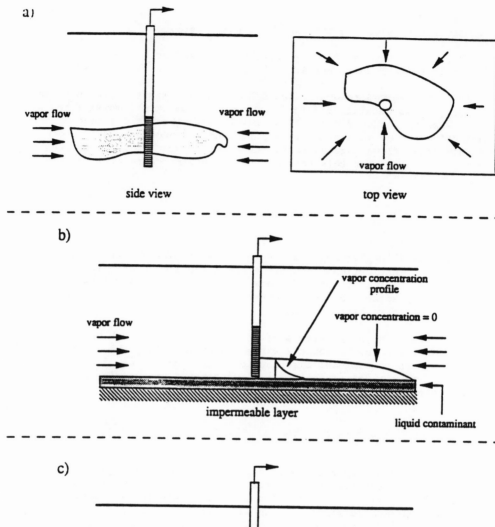

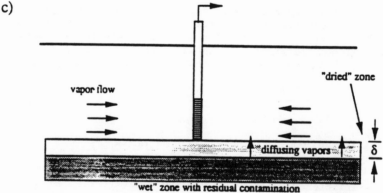

Figure E-6.   Scenarios for Removal Rate Estimates.

P_{Atm} - absolute ambient pressure - 1.016×10^6 g/cm-s^2
P_v - absolute pressure at the venting well [g/cm-s^2]
$R_1 < r < R_2$ - defines region in which contamination is present

Note that the efficiency η is inversely proportional to the screened interval thickness H because a larger interval will, in this geometry, pull in unsaturated air that has passed above the liquid-phase contamination. D is calculated by the Millington-Quirk[3] expression, which utilizes the molecular diffusion coefficient in air D°, the vapor-filled soil porosity ϵ_A, and the total soil porosity ϵ_T:

$$\text{(E-8)}$$

$$D = D° \frac{\epsilon_A^{3.33}}{\epsilon_t^2}$$

where ϵ_A and ϵ_t are related by:

$$\epsilon_A = \epsilon_t - \rho_b \theta_M \qquad \text{(E-9)}$$

Here ρ_b and θ_M are the soil bulk density [g/cm^3] and soil moisture content [g-H$_2$O/g-soil].

As an example, consider removing a layer of contamination bounded by sandy soil (k=1 darcy). A 5.1-cm (2") radius extraction well is being operated at P_v=0.90 atm (0.91×10^6 g/cm-s^2), and the contamination extends from the region $R_1 = R_v = 5.1$ cm to $R_2 = 9$ m (30 ft). The well is screened over a 3m (10 ft) interval. Assuming that:

ρ_b = 1.6 g/cm^3
θ_M = 0.10
D°   = 0.087 cm^2/s
ϵ_T = 0.30
R_I = 12 m

then the venting efficiency relative to the maximum removal rate (Equation E-5), calculated from Equations E-7 through E-9 is:

$$\eta = 0.09 = 9\%$$

Figure E-6c depicts the situation in which vapor flows primarily past, rather than through the contaminated soil zone, such as might be the case for a contaminated clay lens surrounded by sandy soils. In this case vapor phase diffusion through the clay to the flowing vapor limits the removal rate. The maximum removal rate in this case occurs when the vapor flow is fast enough to maintain a very low vapor concentration at the permeable/impermeable soil interface. At any time t a contaminant-free or "dried out" zone of low permeability will exist with a thickness δ. An estimate of the removal rate R_{est} from a contaminated zone extending from R_1 to R_2 is:

$$R_{est} = \pi(R_2^2 - R_1^2)C_{est}D/\delta(t) \qquad (E-10)$$

where D is the effective porous media vapor diffusion coefficient (as calculated above from Equations E-8 and E-9) and C_{est} is the estimated equilibrium vapor concentration (Equation E-1). With time $\delta(t)$ will grow larger. In the case of a single component system the dry zone thickness can be calculated from the mass balance:

$$\qquad (E-11)$$

$$\rho_b C_s \frac{d\delta}{dt} = C_{est}D/\delta(\tau)$$

where C_s is the residual level of contamination in the low permeability zone [g-contamination/g-soil], and all other variables are defined above. The solution to Equations E-10 and E-11 yields the following equation that predicts the change in removal rate with time:

$$\delta(\tau) = \frac{[2C_{est}Dt]^{1/2}}{\rho_b C_s}$$

$$\qquad (E-12)$$

$$R_{est} = \pi(R_2^2 - R_1^2)\frac{[C_{est}DC_s\rho_b]^{1/2}}{2\tau}$$

As an example, consider the case where benzene ($C_v = 3.19 \times 10^{-4}$ g/cm^3 @20°C) is being removed from a zone extending from $R_1 = 5.1$ cm to $R_2 = 9$ m. The initial residual level is 10,000 ppm (0.01 g-benzene/g-soil), $\rho_b = 1.6$ g/cm^3, $D^o = 0.087$ cm^2/s, and $\epsilon_T = \epsilon_A = 0.30$. Figure E-7 presents the predicted removal rates and "dry" zone thickness d(t) as a function of time. Note that it would take approximately one year to clean a layer 1.5 m (5 ft) thick, for a compound as volatile as benzene. Equation E-12 predicts very high initial removal rates; in practice, however, the removal rate will be limited initially by the vapor-phase diffusion behavior described above for Figure E-6b.

Mixture removal rates for the situations depicted in Figures E-6b and E-6c are difficult to estimate because changes in composition and liquid-phase diffusion affect the behavior. Presently there are no simple analytical solutions for these situations, but we can postulate that they should be less than the rates predicted above for pure components.

The use of equilibrium-based models to predict required removal rates is discussed below under the next question.

(5) - What are the vapor composition and concentration changes? What residual, if any, will be left in the soil?

As contaminants are removed during venting, the residual soil contamination level decreases and mixture compositions become richer in the

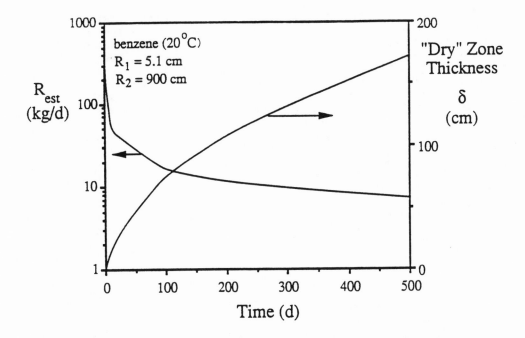

Figure E-7.   Estimated Maximum Removal Rates for a Venting Operation
Limited by Diffusion.

less volatile compounds. Both of these processes result in decreased vapor
concentrations, and hence, decreased removal rates with time. At low
residual soil contamination levels (<500 ppm) Equation E-1 becomes less valid
as sorption and dissolution phenomena begin to affect the soil residual -
vapor equilibrium. In the limit of low residual contamination levels,
contaminant equilibrium vapor concentrations are expected to become
proportional to the residual soil contaminant concentrations. As venting
continues and residual soil levels decrease, therefore, it becomes more
difficult to remove the residual contamination. It is important to realize
that, even with soil venting, there are practical limitations on the final
soil contamination levels that can be achieved. Knowledge of these limits is
necessary to realistically set clean-up criteria and design effective venting
systems.

    The maximum efficiency of a venting operation is limited by the
equilibrium partitioning of contaminants between the soil matrix and vapor
phases. The maximum removal rate is achieved when the vapor being removed
from an extraction well is in equilibrium with the contaminated soil. Models
for predicting this maximum removal rate have been presented by Marley and
Hoag[4] and Johnson et al.[2] The former considered only compositions in a
residual free-phase, while the latter also considered the effects of sorption
and dissolution processes. A complete discussion of the development of these
models is not appropriate here, but we will discuss use of the predictions.

The change in composition, vapor concentration, removal rate, and residual soil contamination level with time are functions of the initial residual composition, vapor extraction well flowrate, and initial soil contamination level. It is not necessary to generate predictions for every combination of variables, however, because with appropriate scaling all results will form a single curve for a given initial mixture composition. Figure E-8a presents the results computed with the model presented by Johnson et al.[2] for the "weathered" gasoline mixture whose composition is given by Table E-2. The important variable that determines residual soil levels, vapor concentrations, and removal rates is the ratio $Qt/M(t=0)$, which represents the volume of air drawn through the contaminated zone per unit mass of contaminant. In Figure E-8, the scaled removal rate (or equivalently the vapor concentration) decreases with time as the mixture becomes richer in the less volatile compounds.

While a detailed compositional analysis was available for this gasoline sample, an approximate composition based on a boiling point distribution curve predicts similar results. Figure E-8b presents the results for the approximate mixture composition also given in Table E-2.

Model predictions, such as those shown in Figure E-8 for the gasoline sample defined by Table E-2, can be used to estimate removal rates (if the vapor flowrate is specified), or alternatively the predictions can be used to estimate vapor flowrate requirements (if the desired removal rate is specified). For example, if we wanted to reduce the initial contamination level by 90%, then Figure E-8 predicts that ≈100 l-air/g-gasoline will be required. This is the minimum amount of vapor required, because it is based on an equilibrium-based model. The necessary minimum average vapor flowrate is then equal to the spill mass times the minimum required vapor flow/mass gasoline divided by the desired duration of venting. Use of this approach is illustrated in the service station site example provided at the end of this paper.

Figure E-8 also illustrates that there is a practical limit to the amount of residual contaminant that can be removed by venting alone. For example, it will take a minimum of 100 l-vapor/g-gasoline to remove 90% of the weathered gasoline defined in Table E-2, while it will take about 200 l-air/g-gasoline to remove the remaining 10%. In the case of gasoline, by the time 90% of the initial residual has been removed the residual consists of relatively insoluble and nonvolatile compounds. It is important to recognize this limitation of venting, and when setting realistic clean-up target levels, they should be based on the potential environmental impact of the residual rather than any specific total residual hydrocarbon levels.

(6) - Are there likely to be any negative effects of soil venting?

It is possible that venting will induce the migration of off-site contaminant vapors towards the extraction wells. This is likely to occur at a service station, which is often in close proximity to other service stations. If this occurs, one could spend a lot of time and money to unknowingly clean-up someone else's problem. The solution is to establish a "vapor barrier" at the perimeter of the contaminated zone. This can be accomplished by allowing vapor flow into any perimeter groundwater monitoring wells, which

TABLE E-2. COMPOSITION OF "FRESH" AND "WEATHERED" GASOLINES

Compound Name	MW (g)	Fresh Gasoline	Weathered Gasoline
propane	44.1	0.0001	0.0000
isobutane	58.1	0.0122	0.0000
n-butane	58.1	0.0629	0.0000
trans-2-butene	56.1	0.0007	0.0000
cis-2-butene	56.1	0.0000	0.0000
3-methyl-1-butene	70.1	0.0006	0.0000
isopentane	72.2	0.1049	0.0069
1-pentene	70.1	0.0000	0.0005
2-methyl-1-butene	70.1	0.0000	0.0008
2-methyl-1,3-butadiene	68.1	0.0000	0.0000
n-pentane	72.2	0.0586	0.0095
trans-2-pentene	70.1	0.0000	0.0017
2-methytl-2-butene	70.1	0.0044	0.0021
3-methyl-1,2-butadiene	68.1	0.0000	0.0010
3,3-dimethyl-1-butene	84.2	0.0049	0.0000
cyclopentane	70.1	0.0000	0.0046
3-methyl-1-pentene	84.2	0.0000	0.0000
2,3-dimethylbutane	86.2	0.0730	0.0044
2-methylpentane	86.2	0.0273 ·	0.0207
3-methylpentane	86.2	0.0000	0.0186
n-hexane	86.2	0.0283	0.0207
methylcyclopentane	84.2	0.0083	0.0234
2,2-dimethylpentane	100.2	0.0076	0.0064
benzene	78.1	0.0076	0.0021
cyclohexane	84.2	0.0000	0.0137
2,3-dimethylpentane	100.2	0.0390	0.0000
3-methylhexane	100.2	0.0000	0.0355
3-ethylpentane	100.2	0.0000	0.0000
n-heptane	100.2	0.0063	0.0447
2,2,4-trimethylpentane	114.2	0.0121	0.0503
methylcyclohexane	98.2	0.0000	0.0393
2,2-dimethylhexane	114.2	0.0055	0.0207
toluene	92.1	0.0550	0.0359
2,3,4-trimethylpentane	114.2	0.0121	0.0000
3-methylheptane	114.2	0.0000	0.0343
2-methylheptane	114.2	0.0155	0.0324
n-octane	114.2	0.0013	0.0300
2,4,4-trimethylhexane	128.3	0.0087	0.0034
2,2-dimethylheptane	128.3	0.0000	0.0226
ethylbenzene	106.2	0.0000	0.0130
p-xylene	106.2	0.0957	0.0151
m-xylene	106.2	0.0000	0.0376
3,3,4-trimethylhexane	128.3	0.0281	0.0056
o-xylene	106.2	0.0000	0.0274
2,2,4-trimethylheptane	142.3	0.0105	0.0012
n-nonane	128.3	0.0000	0.0382
3,3,5-trimethylheptane	142.3	0.0000	0.0000
n-propylbenzene	120.2	0.0841	0.0117
2,3,4-trimethylheptane	142.3	0.0000	0.0000
1,3,5-trimethylbenzene	120.2	0.0411	0.0493
1,2,4-ttrimethylbenzene	120.2	0.0213	0.0705
n-decane	142.3	0.0000	0.0140
methylpropylbenzene	134.2	0.0351	0.0170
dimethylethylbenzene	134.2	0.0307	0.0289
n-undecane	156.3	0.0000	0.0075
1,2,4,5,-tetramethylbenzene	134.2	0.0133	0.0056
1,2,3,4,-tetramethylbenzene	134.2	0.0129	0.0704
1,2,4-trimethyl-5-ethylbenzene	148.2	0.0405	0.0651
n-dodecane	170.3	0.0230	0.0000
naphthalene	128.2	0.0045	0.0076
n-hexylbenzene	162.3	0.0000	0.0147
methylnaphthalene	142.2	0.0023	0.0134
TOTAL		1.0000	1.0000

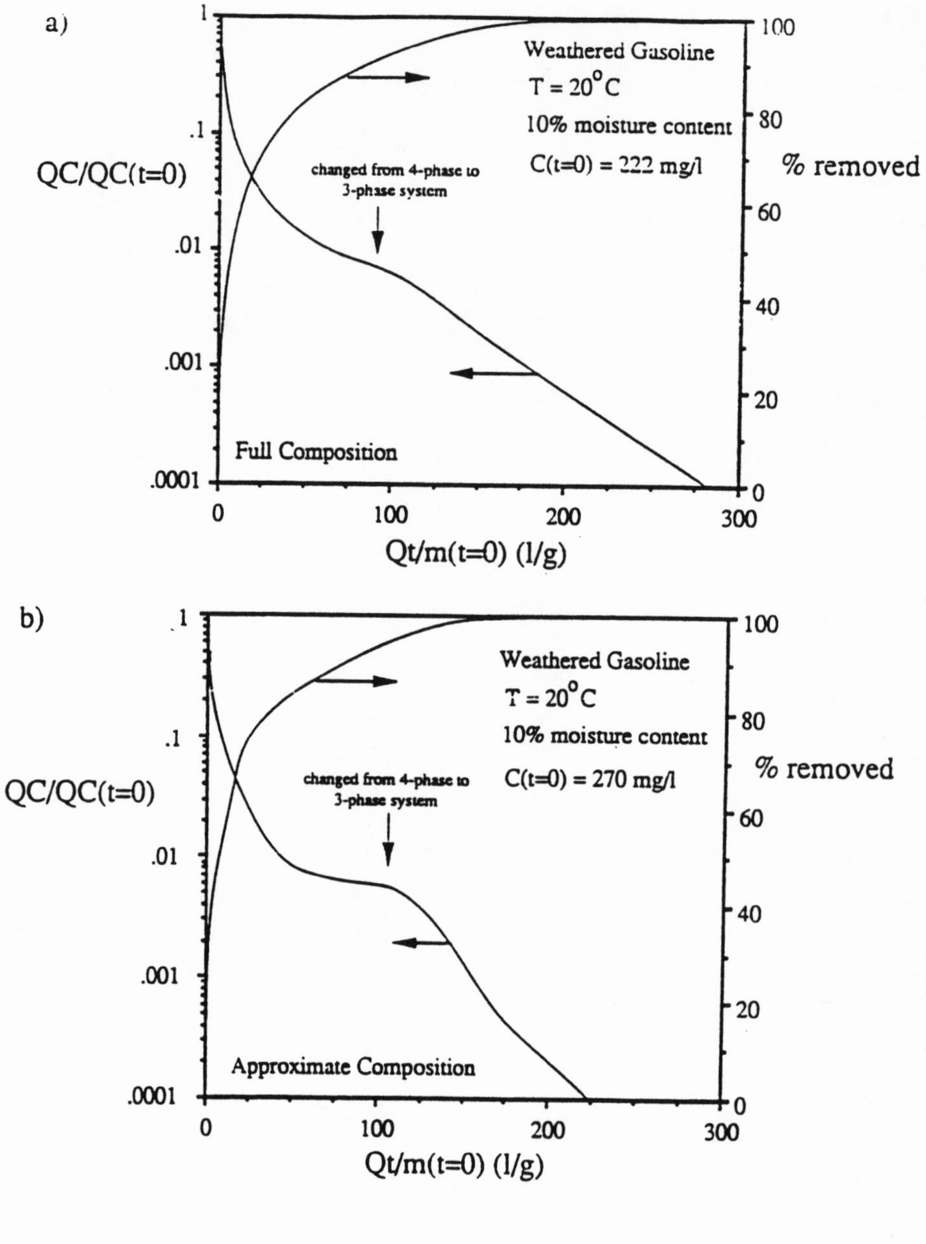
a)
Weathered Gasoline
T = 20°C
10% moisture content
C(t=0) = 222 mg/l
changed from 4-phase to
3-phase system
Full Composition
QC/QC(t=0)
% removed
Qt/m(t=0) (l/g)
1
.1
.01
.001
.0001
0
100
200
300
100
80
60
40
20
0
b)
Weathered Gasoline
T = 20°C
10% moisture content
C(t=0) = 270 mg/l
changed from 4-phase to
3-phase system
Approximate Composition
QC/QC(t=0)
% removed
Qt/m(t=0) (l/g)
1
.1
.01
.001
.0001
0
100
200
300
100
80
60
40
20
0

Figure E-8. Maximum Predicted Removal Rates for a Weathered Gasoline. a) full composition, b) approximate composition.

then act as passive air supply wells. In other cases it may be necessary to install passive air injection wells, or trenches, as illustrated in Figure E-9a.

As pointed out by Johnson et al.[2] the application of a vacuum to extraction wells can also cause a water table rise. In many cases contaminated soils lie just above the water table and they become water saturated, as illustrated in Figure E-9b. The maximum rise occurs at the vapor extraction well, where the water table rise will be equal to the vacuum at the well expressed as an equivalent water column height (i.e., in or ft H_2O). The solution to this problem is to install a dewatering system, with groundwater pumping wells located as close to vapor extraction wells as possible. The dewatering system must be designed to insure that contaminated soils remain exposed to vapor flow. Other considerations not directly related to venting system design, such as soluble plume migration control and free-liquid product yield, will also be factors in the design of groundwater pumping system.

Design Information

If venting is still a remediation option after answering the questions above, then more accurate information must be collected. Specifically, the soil permeability to vapor flow, vapor concentrations, and aquifer characterics are required. These are obtained by two field experiments: air permeability and groundwater pump tests. These are described briefly below.

Air Permeability Tests

Figure E-10 depicts the set-up of an air permeability test. The object of this experiment is to remove vapors at a constant rate from an extraction well, while monitoring with time the transient subsurface pressure distribution at fixed points. Effluent vapor concentrations are also monitored. It is important that the test be conducted properly to obtain accurate design information. The extraction well should be screened through the soil zone that will be vented during the actual operation. In many cases existing groundwater monitoring wells are sufficient, if their screened sections extend above the water table. Subsurface pressure monitoring probes can be driven soil vapor sampling probes (for shallow <20 ft deep contamination problems) or more permanent installations.

Flowrate and transient pressure distribution data are used to estimate the soil permeability to vapor flow. The expected change in the subsurface pressure distribution with time $P'(r,t)$ is predicted[2] by:

$$P' = \frac{Q}{4\pi m(k/\mu)} \int_{\frac{r^2 \epsilon\mu}{4kP_{Atm}t}}^{\infty} \frac{e^{-x}}{x} \, dx \qquad \text{(E-13)}$$

a)

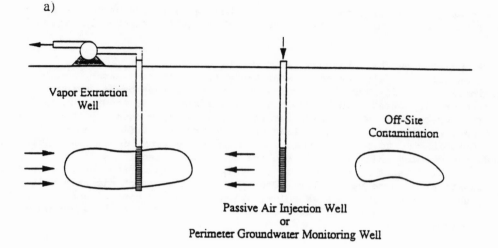

b)

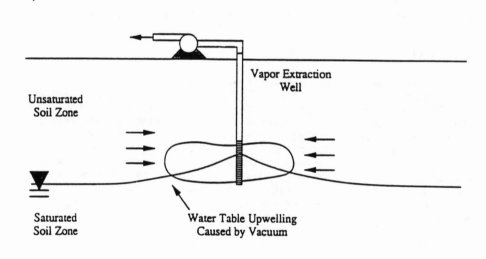

Figure E-9. a) Use of Passive Vapor Wells to Prevent Migration of Off-Site
Contaminant Vapors. b) Water Table Rise Caused by the Applied Vacuum.

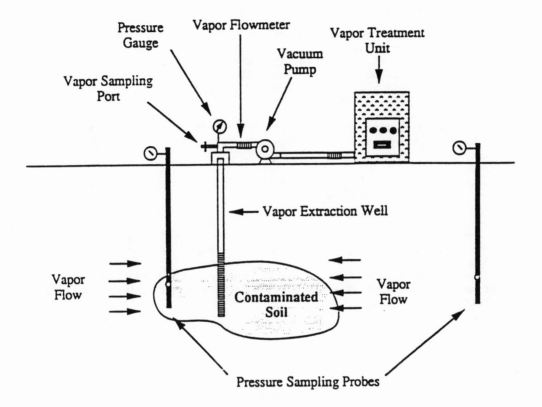

Figure E-10. Air Permeability Test System.

For $(r^2/4kP_{Atm}t)<0.1$,   Equation E-13 can be approximated by:

$$P' = \frac{Q}{4\pi m(k/\mu)}[-0.5772 - \ln(\frac{r^2\epsilon\mu}{4kP_{Atm}}) + \ln(t)] \qquad (E-14)$$

where:

P'	=	"gauge" pressure measured at distance r and time t
m	=	stratum thickness
r	=	radial distance from vapor extraction well
k	=	soil permeability to air flow
μ	=	viscosity of air = 1.8×10^{-4} g/cm-s
ϵ	=	air-filled soil void fraction
t	=	time
Q	=	volumetric vapor flowrate from extraction well
P_{Atm}	=	ambient atmospheric pressure = 1.0 atm = 1.013×10^6 g/cm-s^2

Equation E-14 predicts that a plot of P' -vs- $\ln(t)$ should be a straight line with slope A and y-intercept B equal to:

$$A = \frac{Q}{4\pi m(k/\mu)} \qquad B = \frac{Q}{4\pi m(k/\mu)}[-0.5772 - \ln(\frac{r^2\epsilon\mu}{4kP_{Atm}})] \qquad (E-15)$$

The permeability to vapor flow can then be calculated from the data by one of two methods. The first is applicable when Q and m are known. The calculated slope A is used:

$$k = \frac{Q\mu}{4A\pi m} \qquad (E-16)$$

The second approach must be used whenever Q or m is not known. In this case the values A and B are both used:

$$k = \frac{r^2\epsilon\mu}{4P_{Atm}}\exp(\frac{B}{A} + 0.5772) \qquad (E-17)$$

Equation E-13 can also be used to choose the locations of subsurface pressure monitoring points before conducting the air permeability test, given an estimation of k and the flowrate to be used.

Vapor samples should be taken at the beginning and end of the air permeability test, which should be conducted for a long enough time to extract at least one "pore volume" V_p of vapor from the contaminated soil zone. This insures that all vapors existing in the formation prior to venting are removed. The vapor concentration at the start of the test is representative of the equilibrium vapor concentration, while the concentration measured after one pore volume has been extracted gives an indication of realistic removal rates and the mixing or diffusional limitations discussed in association with

Figure E-6.   The time τ_P for one pore volume to be removed is:

$$\tau_P - V_P/Q - \epsilon_A\pi R^2 H/Q \qquad\qquad (E-18)$$

where R, H, ϵ_A, and Q are the radius of the zone of contamination, vertical thickness of the zone of contamination, air-filled void fraction, and volumetric vapor flowrate from the extraction well. For example, consider the case where R-12 m, H-3 m, ϵ_A-0.35, and Q-0.57 m³/min (20 ft³/min). Then τ_P-475 m³/0.57 m³/min-833 min-14 h.

Groundwater Pump Tests

To achieve efficient venting the hydrocarbon-contaminated soil has to be exposed to air flow, which in turn requires that the water table be lowered to counteract the water upwelling effect caused by the decreased vapor pressure in the vicinity of a venting well (Johnson et al.[2]) and to possibly expose contaminated soil below the water table. Thus the groundwater pumping system has to have a sufficient pumping rate and be operated for a long enough time period to obtain the required drawdowns. Since most venting systems are installed above phreatic aquifers, two aquifer parameters are needed for the design: average transmissivity T and effective porosity S. These parameters can be estimated using the results of the standard transient groundwater pump test with a constant pumping rate (Bear[5]). Using the estimated values the required pumping rate may be calculated as follows:

$$Q - 4\pi TS(r,t)/W(u) \qquad\qquad (E-19)$$

where: W(u) is the well function[5] of $u - Sr^2/4Tt$, and s(r,t) is the required drawdown at distance r and pumping time equal to t.

System Design

In this section we discuss the questions that must be answered in order to design an in-situ soil venting system. It is not our intention to provide a generic "recipe" for soil venting systems design; instead we suggest a structured thought process to guide in choosing the number of extraction wells, well spacing, construction, etc. Even in a structured thought process, intuition and experience play important roles. There is no substitute for a good fundamental understanding of vapor flow processes, transport phenomena, and groundwater flow.

- *Choosing the number of vapor extraction wells*

Three methods for choosing the number of vapor extraction wells are outlined below. The greatest number of wells from these three methods is then the value that should be used. The objective is to satisfy removal rate requirements and achieve vapor removal from the entire zone of contamination.

For the first estimate we neglect residual contaminant composition and vapor concentration changes with time. The acceptable removal rate $R_{acceptable}$ is calculated from Equation E-4, while the estimated removal rate from a single well R_{est} is estimated from a choice of Equations E-2, E-6, E-7, or E-12 depending on whether the specific site conditions are most like Figure E-6a, E-6b, or E-6c. The number of wells N_{well} required to achieve the acceptable removal rate is:

$$N_{well} = R_{acceptable}/R_{est}$$ (E-20)

Equations E-2, E-6, and E-7 require vapor flow estimates, which can be calculated from Equation E-5 using the measured soil permeability and chosen extraction well vacuum P_w. At this point one must determine what blowers and vacuum pumps are available because the characteristics of these units will limit the range of feasible (P_w,Q) values. For example, a blower that can pump 100 scfm at 2 in H_2O vacuum may only be able to pump 10 scfm at 100 in H_2O vacuum.

The second method, which accounts for composition changes with time, utilizes model predictions, such as those illustrated in Figure E-8. Recall that equilibrium-based models are used to calculate the minimum vapor flow to achieve a given degree of remediation. For example, if we wish to obtain a 90% reduction in residual gasoline levels, Figure E-8 indicates that ≈100 l-vapor/g-gasoline must pass through the contaminated soil zone. If our spill mass is 1500 kg (≈500 gal), then a minimum of 1.5×10^8 l-vapor must pass through the contaminated soil zone. If our target clean-up period is six months, this corresponds to a minimum average vapor flowrate of 0.57 m^3/min (≈20 cfm). The minimum number of extraction wells is then equal to the required minimum average flowrate/flowrate per well.

The third method for determining the number of wells insures that we remove vapors and residual soil contamination from the entire zone of contamination N_{min}. This is simply equal to the ratio of the area of contamination $A_{contamination}$, to the area of influence of a single venting well $\pi R_I 2$:

$$N_{min} = \frac{A_{contamination}}{\pi R_I^2}$$ (E-21)

This requires an estimate of R_I, which defines the zone in which vapor flow is induced. In general, R_I depends on soil properties of the vented zone, properties of surrounding soil layers, the depth at which the well is screened, and the presence of any impermeable boundaries (water table, clay layers, surface seal, building basement, etc.). At this point it is useful to have some understanding of vapor flow patterns because, except for certain ideal cases[6], one cannot accurately predict vapor flowpaths without numerically solving vapor flow equations. An estimate for R_I can be obtained by fitting radial pressure distribution data from the air permeability test to the steady-state radial pressure distribution equation[2]:

$$P(r) = P_v[1+(1-(\frac{P_{Atm}}{P_v})^2)\frac{\ln(r/R_v)}{\ln(R_v/R_I)}]^{1/2} \qquad (E-22)$$

where $P(r)$, P_{Atm}, P_v, and R_v are the absolute pressure measured at a distance r from the venting well, absolute ambient pressure, absolute pressure applied at the vapor extraction well, and extraction well radius, respectively. Given that these tests are usually conducted for less than a day, the results will generally underestimate R_I. If no site specific data is available, one can conservatively estimate R_I based on the published reports from in-situ soil venting operations. Reported R_I values for permeable soils (sandy soils) at depths greater than 20 ft below ground surface, or shallower soils beneath good surface seals, are usually 10 m - 40 m.[1] For less permeable soils (silts, clays), or more shallow zones R_I is usually less.

- *Choosing well location, spacing, passive wells, and surface seals*

To be able to successfully locate extraction wells, passive wells, and surface seals one must have a good understanding of vapor flow behavior. We would like to place wells so that we insure adequate vapor flow through the contaminated zone, while minimizing vapor flow through other zones.

If one well is sufficient, it will almost always be placed in the geometric center of the contaminated soil zone, unless it is expected that vapor flow channeling along a preferred direction will occur. In that case the well will be placed so as to maximize air flow through the contaminated zone.

When multiple wells are used it is important to consider the effect that each well has on the vapor flow to all other wells. For example, if three extraction wells are required at a given site, and they are installed in the triplate design shown in Figure E-11a, there would be a "stagnant" region in the middle of the wells where air flow would be very small in comparison to the flow induced outside the triplate pattern boundaries. This problem can be alleviated by the use of "passive wells" or "forced injection" wells as illustrated in Figure E-11b (it can also be minimized by changing the vapor flowrates from each well with time). A passive well is simply a well that is open to the atmosphere; in many cases groundwater monitoring wells are suitable. If a passive or forced injection well is to have any positive effect, it must be located within the extraction well's zone of influence. Forced injection wells are simply vapor wells into which air is pumped rather than removed. One must be very careful in choosing the locations of forced injection wells so that contaminant vapors are captured by the extraction wells, rather than forced off-site. To date there have not been any detailed reports of venting operations designed to study the advantages/disadvantages of using forced injection wells. Figure E-11c presents another possible extraction/injection well combination. As illustrated in Figure E-9, passive wells can also be used as vapor barriers to prevent on-site migration of off-site contamination problems.

For shallow contamination problems (<4 m below ground surface) vapor extraction trenches combined with surface seals may be more effective than

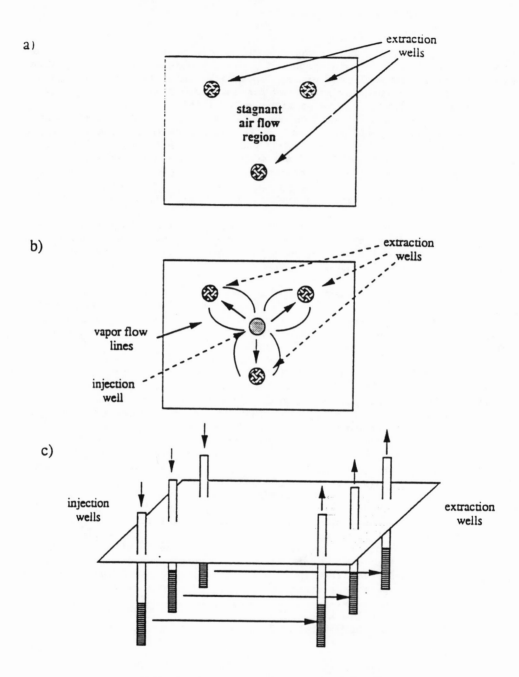

Figure E-11.   Venting Well Configuration.

vertical wells. Trenches are usually limited to shallow soil zones because the difficulty of installation increases with depth.

Surface seals, such as polymer-based liners and asphalt, concrete, or clay caps, are sometimes used to control the vapor flow paths. Figure E-12 illustrates the effect that a surface seal will have on vapor flow patterns. For shallow treatment zones (<5 m) the surface seal will have a significant effect on the vapor flow paths, and seals can be added or removed to achieve the desired vapor flowpath.   For wells screened below 8 m the influence of surface seals becomes less significant.

- Well screening and construction

Wells should be screened only through the zone of contamination, unless the permeability to vapor flow is so low that removal rates would be greater if flow were induced in an adjacent soil layer (see Figure E-6). Removal rate estimates for various mass-transfer limited scenarios can be calculated from Equations E-7 and E-12.

Based on Equation E-5, the flowrate is expected to increase by 15% when the extraction well diameter is increased from 10 cm (4 in) to 20 cm (8 in). This implies that well diameters should be as large as is practically possible.

A typical well as shown in Figure E-13a is constructed from slotted pipe (usually PVC). The slot size and number of slots per inch should be chosen to maximize the open area of the pipe. A filter packing, such as sand or gravel, is placed in the annulus between the borehole and pipe . Vapor extraction wells are similar to groundwater monitoring wells in construction but there is no need to filter vapors before they enter the well. The filter packing, therefore, should be as coarse as possible. Any dust carried by the vapor flow can be removed by an above-ground filter. Bentonite pellets and a cement grout are loaded above the filter packing. It is important that these be properly installed to prevent a vapor flow "short-sircuit". Any groundwater monitoring wells installed near the extraction wells must also be installed with good seals.

- Vapor treatment

Currently there are four main treatment processes available. Each is discussed below.

- vapor combustion units: Vapors are incinerated and destruction efficiencies are typically >95%. A supplemental fuel, such as propane, is added before combustion unless extraction well vapor concentrations are on the order of a few percent by volume. This process becomes less economical as vapor concentrations decrease below $\approx$10,000 ppm$_v$.

- catalytic oxidation units: Vapor streams are heated and then passed over a catalyst bed. Destruction efficiencies are typically >95%. These units are used for vapor concentrations <8000 ppm$_v$. More concentrated vapors can cause catalyst bed temperature excursions and melt-down.

a)

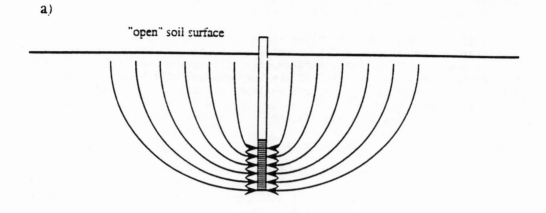

b)

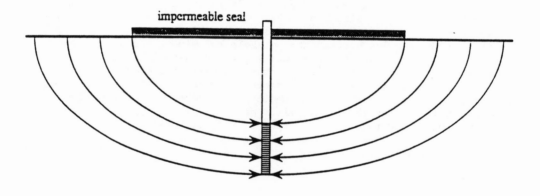

Figure E-12.   Effect of Surface Seal on Vapor Flowpath.

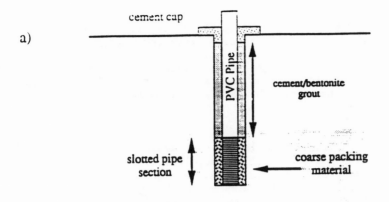

a)

cement cap

PVC Pipe

cement/bentonite grout

slotted pipe section

coarse packing material

b)

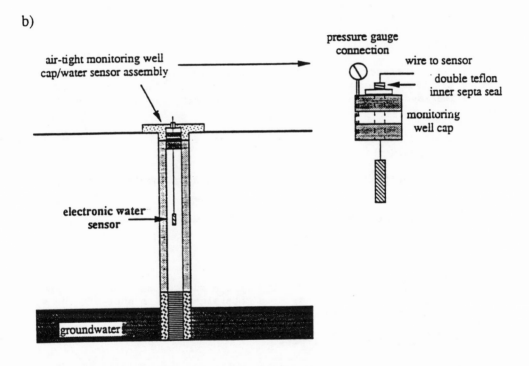

air-tight monitoring well cap/water sensor assembly

electronic water sensor

groundwater

pressure gauge connection

wire to sensor

double teflon inner septa seal

monitoring well cap

Figure E-13.   a) Extraction Well Construction, and b) Air-Tight Groundwater Level Measuring System.

- *carbon beds*: Carbon can be used to treat almost any vapor streams, but is only economical for very low emission rates (<100 g/d)

- *diffuser stacks*: These do not treat vapors, but are the most economical solution for areas in which they are permitted. They must be carefully designed to minimize health risks and maximize safety.

- *Groundwater pumping system*

In cases where contaminated soils lie just above or below the water table, groundwater pumping systems will be required to insure that contaminated soils remain exposed. In designing a groundwater system it is important to be aware that upwelling (draw-up) of the groundwater table will occur when a vacuum is applied at the extraction well (see Figure E-9b). Because the upwelling will be greatest at the extraction wells, groundwater pumping wells should be located within or as close to the extraction wells as possible. Their surface seals must be airtight to prevent unwanted short-circuiting of airflow down the groundwater wells.

- *System integration*

System components (pumps, wells, vapor treating units, etc.) should be combined to allow maximum flexibility of operation. The review by Hutzler et al.[1] provides descriptions of many reported systems. Specific requirements are:

- separate valves, flowmeters, and pressure gauges for each extraction and injection well.

- air filter to remove particulates from vapors upstream of pump and flow meter.

- knock-out pot to remove any liquid from vapor stream upstream of pump and flow meter.

Monitoring

The performance of a soil venting system must be monitored in order to insure efficient operation, and to help determine when to shut-off the system. At a minimum the following should be measured:

- *date and time* of measurement.

- *vapor flow rates* from extraction wells and into injection wells: these can be measured by a variety of flowmeters including pitot tubes, orifice plates, and rotameters. It is important to have calibrated these devices at the field operating pressures and temperatures.

- *pressure readings* at each extraction and injection well can be measured with manometers or magnahelic gauges.

- *vapor concentrations and compositions* from extraction wells: total hydrocarbon concentration can be measured by an on-line total hydrocarbon analyzer calibrated to a specific hydrocarbon. This information is combined with vapor flowrate data to calculate removal rates and the cumulative amount of contaminant removed. In addition, for mixtures the vapor composition should be periodically checked. It is impossible to assess if vapor concentration decreases with time are due to compositional changes or some other phenomena (mass transfer resistance, water table upwelling, pore blockage, etc.) without this information. Vapor samples can be collected in evacuated gas sampling cylinders, stored, and later analyzed.

   - *temperature:* ambient and soil.

   - *water table level* (for contaminated soils located near the water table): It is important to monitor the water table level to insure that contaminated soils remain exposed to vapor flow. Measuring the water table level during venting is not a trivial task because the monitoring well must remain sealed. Uncapping the well releases the vacuum and any effect that it has on the water table level. Figure E-13b illustrates a monitoring well cap (constructed by Applied Geosciences Inc., Tustin, CA) that allows one to measure simultaneously the water table level and vacuum in a monitoring well. It is constructed from a commercially available monitoring well cap and utilizes an electronic water level sensor.

Other valuable, but optional measurements are:

   - *soil gas vapor concentrations and compositions:* these should be measured periodically at different radial distances from the extraction well. Figure E-14 shows the construction of a permanent monitoring installation that can be used for vapor sampling and subsurface temperature measurements. Another alternative for shallow contamination zones is the use of soil gas survey probes.

   This data is valuable for two reasons: a) by comparing extraction well concentrations with soil gas concentrations it is possible to estimate the fraction of vapor that is flowing through the contaminated zone $f=C_{extraction\ well}/C_{soil\ gas}$, and b) it is possible to determine if the zone of contamination is shrinking towards the extraction well, as it should with time. Three measuring points are probably sufficient if one is located near the extraction well, one is placed near the original edge of the zone of contamination, and the third is placed somewhere in between.

   These monitoring installations can also be useful for monitoring the subsurface vapors after venting has ceased.

When To Turn Off The System?

   Target soil clean-up levels are often set on a site-by-site basis, and are based on the estimated potential impact that any residual may have on air quality, groundwater quality, or other health standards. They may also be related to safety considerations (explosive limits). Generally, confirmation soil borings, and sometimes soil vapor surveys, are required before closure is granted. Because these analyses are expensive and often disrupt the normal business of a site, it would be valuable to be able to determine when

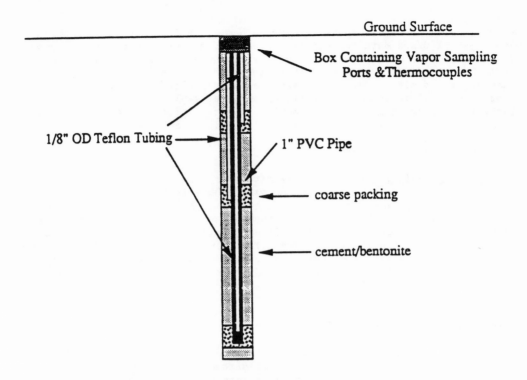

Figure E-14. Vadose Zone Monitoring Installation.

confirmation borings should be taken. If the monitoring is done as suggested above, then the following criteria can be used:

- *cumulative amount removed*: determined by integrating the measured removal rates (flowrate x concentration) with time. While this value indicates how much contaminant has been removed, it is usually not very useful for determining when to take confirmation borings unless the original spill mass is known very accurately. In most cases that information is not available and can not be calculated accurately from soil boring data.

- *extraction well vapor concentrations*: the vapor concentrations are good indications of how effectively the venting system is working, but decreases in vapor extraction well concentrations are not strong evidence that soil concentrations have decreased. Decreases may also be due to other phenomena such as water table level increases, increased mass transfer resistance due to drying, or leaks in the extraction system.

- *extraction well vapor composition*: when combined with vapor concentrations this data gives more insight into the effectiveness of the system. If the total vapor concentration decreases without a change in composition, it is probably due to one of the phenomena mentioned above, and is not an indication that the residual contamination has been significantly reduced. If a decrease in vapor concentration is accompanied by a shift in composition towards less volatile compounds, on the other hand, it is most likely due to a change in the residual contaminant concentration. For residual gasoline clean-up, for example, one might operate a venting system until benzene, toluene, and xylenes were not detected in the vapors. The remaining residual would then be composed of larger molecules, and it can be argued that these do not pose a health threat through volatilization or leaching pathways.

- *soil gas contaminant concentration and composition*: this data is the most useful because it yields information about the residual composition and extent of contamination. Vapor concentrations can not be used to determine the residual level, except for very low residual levels (<500 mg/kg).

<u>Other Factors</u>

- *increased biodegradation*

It is often postulated that because the air supply to the vadose zone is increased, the natural aerobic microbiological activity is increased during venting. While the argument is plausible and some laboratory data is available[7], conclusive evidence supporting this theory has yet to be presented. This is due in part to the difficulty in making such a measurement. A mass balance approach is not likely to be useful because the initial spill mass is generally not known with sufficient accuracy. An indirect method would be to measure CO_2 levels in the extraction well vapors, but this in itself does not rule out the possibility that O_2 is converted to CO_2 before the vapors pass through the contaminated soil zone. The best approach is to measure the O_2/CO_2 concentrations in the vapors at the edge of the contaminated zone, and in the vapor extraction wells. If the CO_2/O_2 concentration ratio increases as the vapors pass through the contaminated soil, one can surmise that a transformation is occurring, although other possible mechanisms (inorganic reactions) must be considered. An increase in aerobic microbial populations would be additional supporting evidence.

- *in-situ heating/venting*

The main property of a compound that determines whether or not it can be removed by venting is its vapor pressure, which increases with increasing temperature. Compounds that are considered nonvolatile, therefore, can be removed by venting if the contaminated soil is heated to the proper temperature. In-situ heating/venting systems utilizing radio-frequency heating and conduction heating are currently under study[8]. An alternative is to reinject heated vapors from catalytic oxidation or combustion units into the contaminated soil zone.

- *air sparging*

Due to seasonal groundwater level fluctuations, contaminants sometimes become trapped below the water table. In some cases groundwater pumping can lower the water table enough to expose this zone, but in other cases this is not practical. One possible solution is to install air sparging wells and then inject air below the water table. Vapor extraction wells would then capture the vapors that bubbled up through the groundwater. To date, success of this approach has yet to be demonstrated. This could have a negative effect if foaming, formation plugging, or downward migration of the residual occurred.

Application of the Design Approach to a Service Station Remediation

In the following we will demonstrate the use of the approach discussed above and outlined in Figure E-2 for the design operation, and monitoring of an in-situ venting operation at a service station.

Preliminary Site Investigation

Prior to sampling it was estimated that 2000 gal of gasoline had leaked from a product line at this site. Several soil borings were drilled and the soil samples were analyzed for total petroleum hydrocarbons (TPH) and other specific compounds (benzene, toluene, xylenes) by a heated-headspace method utilizing a field GC-FID. Figure E-15 summarizes some of the results for one transect at this site. The following relevant information was collected:

- based on boring logs there are four distinct soil layers at this site between 0 - 18 m (0- 60 ft) below ground surface (BGS). Figure E-15 indicates the soil type and location of each of these layers.

- depth to groundwater was 15 m, with fine to medium sand aquifer soils

- the largest concentrations of hydrocarbons were detected in the sandy and silty clay layers adjacent to the water table. Some residual was detected below the water table. Based on the data presented in Figure E-15 it is estimated that ~ 4000 kg of hydrocarbons are present in the lower two soil zones.

- initially there was some free-liquid gasoline floating on the water table, and this was subsequently removed by pumping. A sample of this product was analyzed and its approximate composition (~20% of the compounds could not be identified) is listed in Table E-2 as the "weathered gasoline". The corresponding boiling point distribution curve for this mixture has been presented in Figure E-3.

- vadose zone monitoring installations similar to the one pictured in Figure E-14 were installed during the preliminary site investigation.

Deciding if Venting is Appropriate

For the remainder of the analysis we will focus on the contaminated soils located just above the water table.

- *What contaminant vapor concentrations are likely to be obtained?*

Based on the composition given in Table E-2, and using Equation E-1, the predicted saturated TPH vapor concentration for this gasoline is:

$$C_{est} - 220 \text{ mg/l}$$

Using the "approximate" composition listed in Table E-2 yields a value of 270 mg/l. The measured soil vapor concentration obtained from the vadose zone monitoring well was 240 mg/l. Due to composition changes with time, this will be the maximum concentration obtained during venting.

- *Under ideal flow conditions is this concentration great enough to yield acceptable removal rates?*

Equation E-4 was used to calculate $R_{acceptable}$. Assuming $M_{spill} - 4000$ kg and $t - 180$ d, then:

$$R_{acceptable} - 22 \text{ kg/d}$$

Using Equation E-2, $C_{est} - 240$ mg/l, and $Q - 2800$ l/min (100 cfm):

$$R_{est} - 970 \text{ kg/d}$$

which is greater than $R_{acceptable}$.

- *What range of vapor flowrates can realistically be achieved?*

Based on boring logs the contaminated zone just above the water table is composed of fine to medium sands, which have an estimated permeability $1 < k < 10$ darcy. Using Figure E-5, or Equation E-5, the predicted flowrates for an extraction well vacuum $P_v - 0.90$ atm are:

$0.04 < Q < 0.4$ m^3/m-min $R_v - 5.1$ cm, $R_I - 12$ m
$0.43 < Q < 4.3$ ft^3/ft-min $\quad R_v - 2.0$ in, $R_I - 40$ ft

The thickness of this zone and probable screen thickness of an extraction well is about 2 m (6.6 ft). The total flowrate per well through this zone is estimated to be $0.08 < Q < 0.8$ m^3/min (2.8 cfm $< Q < 28$ cfm).

- *Will the contaminant concentrations and estimated flowrates produce acceptable removal rates?*

Using $C_{est} - 240$ mg/l, the maximum removal rates likely to be obtained are calculated from Equation E-2:

$$28 \text{ kg/d} < (R_{est})_{max} < 280 \text{ kg/d}$$

To be conservative, we will guess that only 50% of the vapor actually flows through contaminated soils, so our estimated removal rate per well will be half of these values. The estimated acceptable removal rate $R_{acceptable} - 22$ kg/d falls within this range. Of course this calculation did not take into account the possibility of vapor concentration decreases during venting. We shall take this into account in the next subsection.

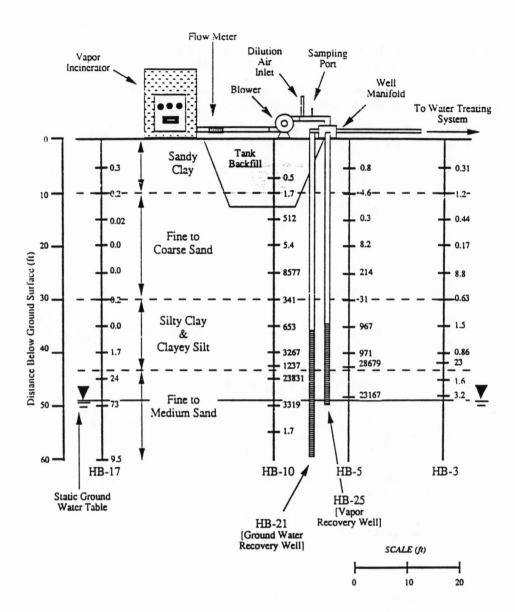

Figure E-15.   Initial Total Hydrocarbon Distribution [mg/kg-soil] and
Location of Lower Zone Vent Well.

- What residual, if any, will be left in the soil?

A target clean-up level for most gasoline spill sites is <1000 mg/kg TPH residual. If our initial residual level is ~10,000 ppm, then we need to remove at least 90% of the initial residual. According to the curves in Figure E-8, which represent the maximum removal rates for the gasoline analyzed at this site, approximately 100 1-vapor/g-residual will have to pass through the contaminated zone to achieve this target. Based on our estimated initial residual of 4000 kg TPH, 4×10^8 1-vapor are required. Over a six month period this corresponds to an average flowrate $Q=1.5$ m^3/min (54 cfm). Recall that since this corresponds to the maximum removal rate, it is the minimum required flowrate.

- Are there likely to be any negative effects of soil venting?

Given that the contaminated soils are located just above and below the water table, water table upwelling during venting must be considered here.

Air Permeability Test

Figure E-16 presents data obtained from the air permeability test of this soil zone. In addition to vapor extraction tests, air injection tests were conducted. The data is analyzed in the same manner as discussed for vapor extraction tests. Accurate flowrate (Q) values were not measured, therefore, Equation E-17 was used to determine the permeability to vapor flow. The k values ranged from 2 to 280 darcys, with the median being ~8 darcys.

System Design

- Number of vapor extraction wells:

Based on the 8 darcys permeability, and assuming a 15 cm diameter (6 in) venting well, a 2 m screened section, $P_w = 0.90$ atm (41 in H_2O vacuum) and $R_I=12$ m, then Equation E-5 predicts:

$$Q = 0.7 \ m^3/min = 25 \ cfm$$

Based on the discussion above, a minimum average flowrate of 1.5 m^3/min is needed to reduce the residual to 1000 ppm in 6 months. The number of wells required is then 1.5/0.7 = 2, assuming that 100% of the vapor flows through contaminated soils. It is not likely that this will occur, and a more conservative estimate of 50% vapor flowing through contaminated soils would require that twice as many wells (4) be installed.

A single vapor extraction well (HB-25) was installed in this soil layer with the knowledge that more wells were likely to be required. Its location and screened interval are shown in Figure E-15. Other wells were installed in the clay layer and upper sandy zone, but in this paper we will only discuss results from treatment of the lower contaminated zone. A groundwater pumping well was installed to maintain a 2 m drawdown below the static water level. Its location is also shown in Figure E-15.

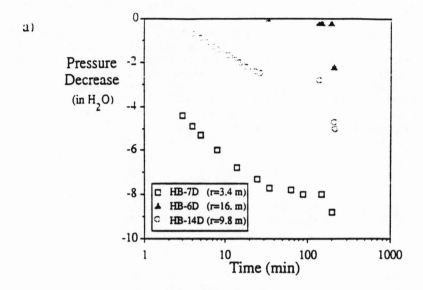

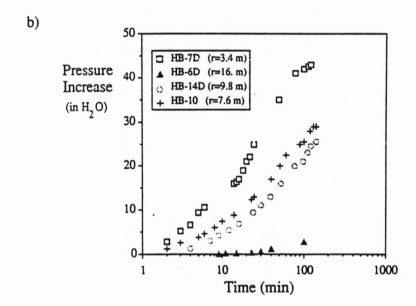

Figure E-16. Air Permeability Test Results: a) vapor extraction test, b) air injection test.

System Monitoring

Three vadose monitoring wells similar in construction to the one pictured in Figure E-14 were installed so that the soil temperature, soil gas concentrations, and subsurface pressure distribution could be monitored at three depths. One sampling port is located in the zone adjacent to the aquifer. The vapor flowrate from HB-25 and vapor concentrations were measured frequently, and the vapor composition was determined by GC-FID analysis. In addition, the water level in the groundwater monitoring wells was measured with the system pictured in Figure E-13b. The results from the first four months of operation are discussed below.

In Figure E-17a the extraction well vacuum and corresponding vapor flowrate are presented. The vacuum was maintained at 0.95 atm (20 in H_2O vacuum), and the flowrate was initially 12 scfm. It gradually decreased to about 6 scfm over 80 d. For comparison, Equation (5) predicts that Q=12 cfm for k=8 darcys. Increasing the applied vacuum to 0.70 atm (120 in H_2O vacuum) had little effect on the flowrate. This could be explained by increased water table upwelling, which would act to decrease the vertical cross-section available for vapor flow. The scatter in the flowrate measurements is probably due to inconsistent operation of the groundwater pumping operation, which frequently failed to perform properly.

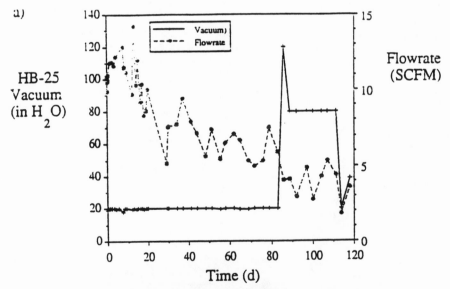

[in H_2O] denote vacuums expressed as equivalent water column heights.

Figure E-17. Soil Venting Results: a) Vacuum/Flowrate Data,

Figure E-17b presents the change in vapor concentration with time. Fifteen specific compounds were identified during the GC-FID vapor analyses; in this figure we present the total concentration of known and unknown compounds detected between five boiling point ranges:

methane - isopentane (<28°C)
isopentane - benzene (28 - 80°C)
benzene - toluene (80 - 111°C)
toluene - xylenes (111 - 144°C)
>xylenes (>144°C)

There was a shift in composition towards less volatile compounds in the first 20 d, but after that period the composition remained relatively constant. Note that there is still a significant fraction of volatile compounds present. Within the first two days the vapor concentration decreased by 50%, which corresponds to the time period for the removal of the first pore volume of air. Comparing the subsequent vapor concentrations with the concentrations measured in the vadose zone monitoring wells indicates that only (80 mg/l)/(240 mg/l)*100=33% of the vapors are flowing through contaminated soil.

Figure E-18a presents calculated removal rates (flowrate x concentration) and cumulative amount (1 gal = 3 kg) removed during the first four months. The decrease in removal rate with time is due to a combination of decreases in flowrate and hydrocarbon vapor concentrations. After the first four months approximately one-fourth of the estimated residual has been removed from this lower zone.

On day 80 the vacuum was increased from 20 - 120 in H$_2$O vacuum and the subsequent increase in subsurface vacuum and water table upwelling was monitored. Figure E-18b presents the results. Note that the water table rise paralleled the vacuum increase, although the water table did not rise the same amount that the vacuum did.

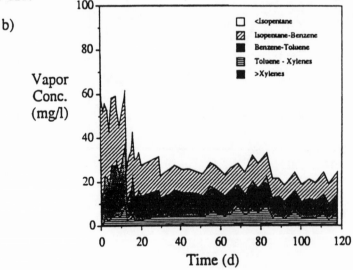

Figure E-17b. Concentration/Composition Data.

a)

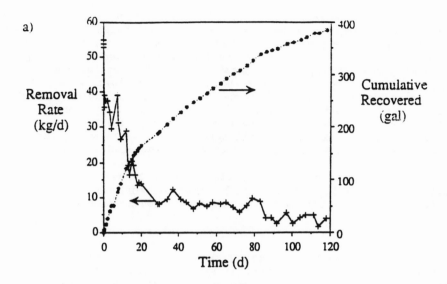

b)

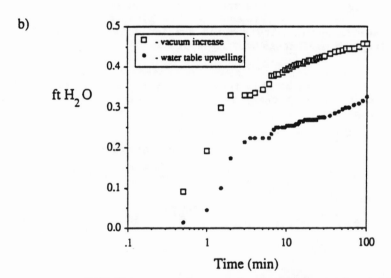

[ft H$_2$O] denote vacuums expressed as equivalent water column heights

Figure E-18. Soil Venting Results: a) Removal Rate/Cumulative
Recovered, b) Water Table Rise.

Figure E-19 compares the reduced measured TPH vapor concentration $C(t)/C(t=0)$ with model predictions. $C(t=0)$ was taken to be the vapor concentration after one pore volume of air had passed through the contaminated zone (~80 mg/l), $m(t=0)$ is equal to the estimated spill mass (~4000 kg), and $V(t)$ is the total volume of air that has passed through the contaminated zone. This quantity is obtained by integrating the total vapor flowrate with time, then multiplying it by the fraction of vapors passing through the contaminated zone f (~0.33). As discussed, the quantity f was estimated by comparing soil gas concentrations from the vadose zone monitoring installations with vapor concentrations in the extraction well vapors. As can be seen, there is good quantitative agreement between the measured and predicted values.

Based on the data presented in Figures E-15 through E-19 and the model predictions in Figure E-8, it appears that more extraction wells (~10 more) are needed to remediate the site within a reasonable amount of time.

CONCLUSIONS

A structured, technically based approach has been presented for the design, construction, and operation of venting systems. While we have attempted to explain the process in detail for those not familiar with venting operations or the underlying governing phenomena, the most effective and efficient systems can only be designed and operated by personnel with a good understanding of the fundamental processes involved. The service station spill example presented supports the validity and usefulness of this approach.

There are still many technical issues that need to be resolved in the future. In particular, we must be able to estimate removal rates for non-ideal situations, demonstrate that biodegradation is enhanced by venting, and investigate novel ideas for enhancing venting removal rates.

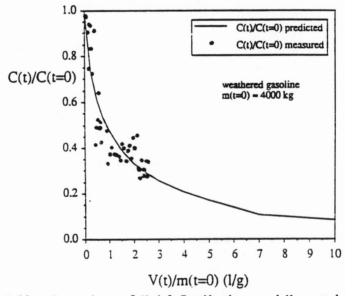

Figure E-19. Comparison of Model Predictions and Measured Response.

REFERENCES

Bear, J., Hydraulics of Groundwater, McGraw-Hill, 1979.

Dev, H., G. C. Sresty, J. E. Bridges and D. Downey, Field Test of the Radio Frequency in-situ Soil Decontamination Process, in Superfund '88: Proceedings of the 9th National Conference, HMCRI, November 1988.

Hutzler, N. J., B. E. Murphy, and J. S. Gierke, State of Technology Review: Soil Vapor Extraction Systems, U.S.E.P.A., CR-814319-01-1, 1988.

Johnson, P. C., M. W. Kemblowski, and J. D. Colthart, Practical Screening Models for Soil Venting Applications, NWWA/API Conference on Petroleum Hydrocarbons and Organic Chemicals in Groundwater, Houston, TX, 1988.

Marley, M.C., and G. E. Hoag, Induced Soil Venting for the Recovery/Restoration of Gasoline Hydrocarbons in the Vadose Zone, NWWA/API Conference on Petroleum Hydrocarbons and Organic Chemicals in Groundwater, Houston, TX, 1984.

Millington, R. J., and J. M. Quirk, Permeability of Porous Solids, Trans. Faraday Soc., 57:1200-1207, 1961.

Salanitro, J. P., M. M. Western, and M. W. Kemblowski, Biodegradation of Aromatic Hydrocarbons in Unsaturated Soil Microcosms. Poster paper presented at the Fourth National Conference on Petroleum Contaminated Soils, University of Massachusetts, Amherst, September 25-28, 1989.

Wilson, D. J., A. N. Clarke, and J. H. Clarke, Soil Clean-up by In-situ Aeration. I. Mathematical Modelling, Sep. Science Tech., 23:991-1037, 1988.

Appendix F
Design of Soil Vapor Extraction Systems—
A Scientific Approach

Michael C. Marley, Scott D. Richter, Bruce L. Cliff, P.E.,
and Peter E. Nangeroni[a]

INTRODUCTION

The fate and transport of volatile, semi-volatile, and gaseous
contaminants released into the sub-surface environment, whether accidentally
or intentionally, has become a subject of primary concern this decade. It has
been estimated that up to 20 percent of the approximately 2 million federally
regulated underground storage tanks in the United States may be leaking.[1]
Spilled product migrates through the unsaturated soil zone, under the
influence of gravitational and capillary forces, to the water table.
Corrective action generally includes an effort to physically remove the
product by bailing and pumping as well as pumping and treating contaminated
ground water. The product retained in the unsaturated zone, however, usually
is a significant portion of the total spill. Natural transport models have
demonstrated the potential of long-term groundwater contamination due to vapor
and solute transport emanating from the trapped immiscible plume.[2,3]

In the past few years the need to remediate these contaminated
unsaturated soils as part of an overall and cost effective site clean-up has
been emphasized. There are a number of methodologies commonly utilized in the
remediation of contaminated unsaturated soils including:

- excavation and off site disposal
- excavation and on-site treatment
- biodegredation
- *in-situ* soil washing
- *in-situ* vapor excavation (soil venting, air stripping, enhanced
  volatilization)

In general, it is recognized that where applicable, vapor extraction is
the most cost effective alternative.[4-10]

[a]VAPEX Environmental Technologies, Inc.
480 Neponset Street
Canton, MA 02021

The cost effectiveness of utilizing vapor extraction has been diminished by somewhat less than optimal employment of the technology. Often, vapor extraction systems have been designed and implemented based on less than a full understanding of the physical/chemical principles governing the process. The focus of this paper is to present a more scientific approach to the design and implementation of vapor extraction systems. This scientific approach is based on several years of research, including the development and application of computer transport models and field sampling and analysis protocols specific to the design, implementation, and understanding of vapor extraction technology.[2,5,6,11-15] The air flow models referred to in the text have been developed as part of the Ph.D. dissertation of one of the co-authors.

VAPOR EXTRACTION TECHNOLOGY

The volatility of the released substances establish a premise for contaminant removal based on inducing air movement in the unsaturated zone. In the unsaturated zone, an air flow field can be established with combinations of injecting and withdrawing boreholes or trenches. Air laden with contaminant vapors would move along the induced flow path toward the withdrawing system where it is analyzed, treated, and/or released to the atmosphere. The remediation of contaminated soils by vapor extraction/soil venting in the past few years has demonstrated the potential effectiveness and economics of this technology.[4-10]

A SCIENTIFIC APPROACH TO DESIGN IMPLEMENTATION

Site Assessment -- Investigation

As with any remediation project, it is important to fully characterize a site in order to develop an optimal remedial approach. On vapor extraction projects, it is important to focus investigative effort on the unsaturated zone. A review of available data on site history and conditions, including site plans (surface and subsurface structures), drilling logs, soils and ground water quality data, potential ground water and vapor receptors, and any site drawings locating utilities. Existing physical and chemical soil analyses will be examined to help evaluate the variability of soil conditions.

Shallow soil gas surveys are effective in providing a preliminary characterization of the degree and extent of soil contamination. Where geologic units, confining to natural vapor transport, prevent contaminant vapors from reaching the capture zone of the shallow soil gas survey, it is necessary to perform a deep soil gas survey. It is important to characterize the contaminant within each separate geologic unit.

Upon data review and analysis, a risk assessment is conducted for the site. Based on the results of a risk assessment, a course of action will be recommended, which, in the majority of cases, constitutes an initiation of the necessary regulatory permitting processes in conjunction with a Phase 2 feasibility study -- conceptual design.

Feasibility Study -- Conceptual Design

A client's attraction for utilizing vapor extraction to remove VOC's from the subsurface is due to the relatively low cost of implementing the technology in conjunction with the higher contaminant removal rates achievable in comparison to standard pump and treat techniques. The high removal rates achievable are related to the properties of the soil matrix, the advective air phase and the physical/chemical properties of the contaminants. The success of the method depends on the rate of contaminant transfer from the immiscible and water phases into the air phase and, in particular, the ability to establish an air flow field that intersects the distributed contaminants.

General Design Approaches --

A more general approach to vapor extraction system design follows the pattern where, the information obtained from the evaluation of the degree and extent of contamination is utilized to provide assumptions of uniform contamination over a specified soil zone. Intraphase equilibrium is assumed between the contaminants and the air and water phases. Uniform air flow fields are assumed and extrapolations of the remediation process are performed. However, following immiscible fluid flow in porous media, the remaining, immobilized, immiscible fluids may exist as a few large globs of liquid, or a large number of smaller globs.[14] The geometry of the fluid distribution depends on the nature of the capillary forces between the fluids, the pore sizes and geometry, and the history of fluid movement in the medium. Although intraphase equilibrium may exist at the pore scale, the heterogeneous distribution of the immobilized immiscible organics within the pores may make the overall equilibrium assumptions inappropriate. It should be noted that where a uniform distribution of residual contamination does exist, the assumption of a dynamic equilibrium between the advective air phase and the immiscible contaminant provides a good approximation of the physical/chemical processes.[5,6,11] The intraphase transfer of contaminants should be considered in terms of mass transfer limitations. At this time, few utilize transport models which consider the potential mass transfer limitation.[3]

However, of greater importance than the potential mass transfer limitations/equilibrium assumptions is a knowledge of and capability to control the airflow pathways to optimize contact with the contaminants. Without the aid of air flow models, it would be difficult to evaluate the air flow pathways for all but the most simplistic of cases (homogeneous, dry sands, closed to the atmosphere, with no subsurface structures within the extraction system zone of influence).

Air Permeability and Air Flow Modeling --

Compressible flow in porous media has been a subject of investigation for many years in petroleum reservoir engineering. Mathematical models of air movement in unsaturated porous media have been calibrated with air pressure data in previous investigations to provide determinations of in-situ air permeability. Muskat and Botset (1931) developed a one-dimensional (radial) air flow model to evaluate the horizontal permeability of gas reservoirs.[16] Boardman and Skrove (1966) injected air into packed-off sections of drill holes and observed radial pressure distributions to obtain horizontal fracture permeability of a granitic rock mass.[17] Stallman and Weeks (1969) and Weeks (1977) describe the use of depth dependent air pressure to calculate vertical air permeability in the unsaturated zone.[18,19] Rozsa and others (1975) document an application of this technique to determine vertical air permeabilities of nuclear chimneys at the Nevada Test Site.[20] As another historical note, soil scientists have utilized injected air and pressure measurements to evaluate soil permeability but these techniques provide estimates over small regions of soil and are not directly applicable for unsaturated zone evaluation.[21-24]

The application of such models to aid in the design of a vapor extraction system is exactly analogous and can be thought of in two steps:

- Evaluate, *in-situ*, the air permeability tensor for the contaminated unsaturated zone by calibrating a steady-state air flow model with pressure measurements obtained during pumping tests.

- Utilize the air permeability values and a steady state air-flow model to determine the well spacings, screened intervals of wells in the unsaturated zone and the size and type of pumps required to generate the desired air movement.

An *in-situ* determination of the air-phase permeability tensor is preferred over laboratory determinations to account for variations in prevailing soil-water conditions, the presence of the immiscible organic liquid, and anisotropy and heterogeneity in air phase permeability. Further, permeability evaluations are sensitive to the soils' bulk density and structure, which are generally altered in the disturbed soil samples taken for laboratory analysis. Steady-state pumping tests, which require less data than transient analyses, are sufficient for this application because only the *in-situ* air permeability is needed for design purposes.

The governing equation defining conservation of, mass for compressible flow is given as:

$$\frac{\partial (\theta_a)}{\partial t} + \nabla \cdot (\rho_a \underline{q_a}) = 0 \tag{1}$$

where

ρ_a = air density

θ_a = air filled porosity

$\underline{q_a}$ = specific discharge vector

Expressing density as a function of pressure in accordance with the ideal gas law and by Darcy's law:

$$\rho_a = \frac{P_a W_a}{R T} \qquad\qquad \underline{q_a} = \frac{\underline{k_a}}{\mu_a} \nabla P \tag{2}$$

where

P_a = air pressure
W_a = molecular weight of air
R = universal gas constant

> T - temperature
> k - intrinsic permeability tensor
> μ_a - air viscosity

yields a partial differential equation in terms of air phase pressure. The selection of a coordinate system and appropriate boundary conditions, together with equation (1) defines the air-flow model.

Commonly, hydraulic conductivity values are available from ground water studies performed prior to the vapor extraction feasibility study. Where applicable (e.g., uniform, dry, medium-coarse, sands), these values may provide an accurate evaluation of the intrinsic horizontal air permeability that could be used in the design process. In the majority of cases, however, this assumption is invalid for one or more of the following reasons:

- gas slippage (Klinkenberg effect [25]) is ignored,
- anisotropy is neglected,
- swelling soils are present,
- the variable water saturation in the unsaturated zone is ignored,
- the presence of an oil phase is ignored,
- the groundwater test may be in a different strata,
- the scale of the ground water test may invalidate the parameter evaluations.

Field Permeability Test

Air Flow System --

A one to two day *in-situ* evaluation of the air permeability tensor in each soil strata of concern is typically performed. Additionally, the data collected will allow a characterization of the surface boundary condition, an important parameter in the development and control of the airflow pathways. The field evaluation normally consists of the installation of one vapor extraction well and several permanent installation vapor probes. The wells and probes installed as part of the feasibility study are located specifically, as to become an integral and efficient part of the overall full scale remediation design. In a number of cases, this single well installation has proven sufficient to effect full scale site remediation. The permanently installed vapor probes allow, over the entire course of the project, for the collection of data on the level of VOC's in the soil gas and hence, to properly document the progress of the cleanup.

The vapor extraction well and vapor probes are installed using standard hollow stem auger techniques. Split spoon sampling is performed to provide soil samples for characterization, initial jar headspace screening and for laboratory analysis by EPA Series 8000 tests. The split spoon sampling is also necessary to detect changes in the stratigraphy that can have significant impact on the developed air flow pathways. This is clearly demonstrated in a field parameter evaluation by Baehr and Hult (1988).[26]

Further, soil samples may be taken to evaluate moisture contents and to develop a moisture content profile. The moisture content data may be used in conjunction with the air permeability evaluations as input data to fit a Corey-Brooks (1966), Parker et al. (1987) or equivalent parametric model to allow interpolation of phase relative permeability-saturation relationships.[27,28] The parametric model may then be used as part of numerical air flow models, in the development of the air permeability grid.

In the analysis of the field parameters, the physical characteristics of the site and the field test layout are used as inputs to the air flow models. The field test is operated at two or more air flow rates; this allows for both the initial calibration of the model (i.e., parameter evaluations using the collected field data at one air flow rate) and verification of the model (i.e., the model is set to simulate the system for the other air flow rates using the parameters established in the calibration made, and comparison is made between the predicted air flow rates and pressure distributions at the well/probes by the model, and the actual pressure data measured at the well/probes at the other air flow rates). Figures F-1 to F-3 demonstrate model calibration, verification and simulation runs for an actual field test where the site geometry and stratigraphy were relatively simplistic. Following calibration/verification, the air flow models are then utilized to establish the optimal, site specific vapor extraction system design, based on achieving the desired airflow rates and pathways.

The operation of a vapor extraction system is accompanied by water movement in the subsurface through:

- Vaporization/condensation to and from the advective air phase;
- The simultaneous water flux in the unsaturated zone induced by the advective air flow; and
- The local ground water mounding due to the negative pressure created within the zone of influence of the vapor extraction well.

Where appropriate, transport models are utilized to evaluate the water movement to ensure a complete extraction system design. In addition to the data collected for the air flow models during the field test, samples are collected and analyzed from the vapor probes and the extraction system discharge gas to better define the degree and extent of contamination and to allow an accurate evaluation of the optimal off-gas treatment system needed in the full scale design. Field technicians and field calibrated instruments (including a portable gas chromatograph/PC system) are used to obtain an *in-situ* qualitative and quantitative evaluation of the contaminants present.

The selection of off-gas controls (required by most states when exhaust gas contamination exceeds a certain level) is based on estimates of the mass and type of contamination in the subsurface. Using the data generated from an air permeability test and determining the minimum airflow required to clean up the zone of contamination, it is possible to determine whether emission controls will be required at a specific site. If possible, design a low flow extraction system that eliminates the need for emission controls. If controls are necessary, either carbon adsorption, catalytic incineration, or thermal incineration will be recommended.

Laboratory Vent Test --

To aid in the evaluation of the feasibility of applying vapor extraction technology and time to clean up, and in consideration of the above discussion

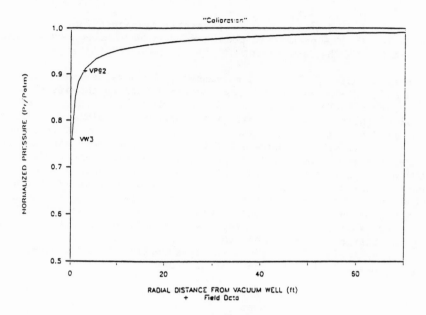

Figure F-1.   Model Simulation for VW3 @ 15 cfm.

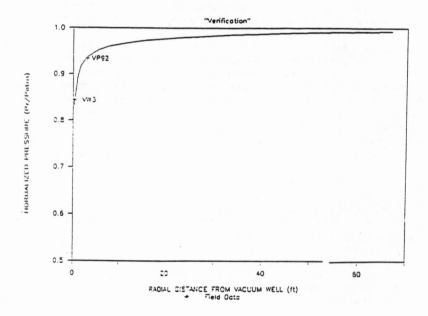

Figure F-2. Model Simulation for VW3 @ 10.5 cfm.

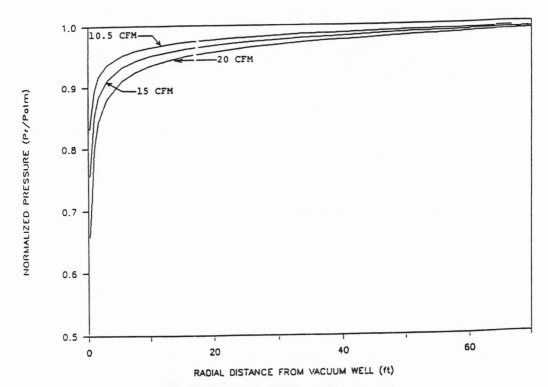

Figure F-3. Model Simulations for VW3.
Simulations @ 15, 10.5, and 20 cfm.

on the mass transfer limitation/equilibrium assumptions, a laboratory vapor
extraction test may be performed on a 2 to 3 kg soil sample, commonly obtained
from the highest contaminated area of the site. The laboratory vent test,
which is essentially a controlled, accelerated vapor extraction process,
provides data on:

- the expected level of clean up achievable utilizing vapor
  extraction technology.
- a quantitative and qualitative (GC/MS) evaluation of the
  contaminants present in the soil system.
- An approximation of the composite intrinsic air
  permeability.
- An estimate of the air flow volume necessary to achieve
  the regulatory required clean up levels imposed on the
  site, and hence in conjunction with the air flow
  modelling an estimate of the time required for site
  remediation.

Mobilization and Completion of Installation

Following receipt of regulatory approval, optimally-sized vacuum
pump(s)/blower(s), manifold piping, emission control (if necessary), and
additional wells and probes (if necessary) are installed.

System Operation, Maintenance and Monitoring

Regular monitoring of the system is recommended to insure its effectiveness and to document and optimize the progress of cleanup. Vapor samples and air flow readings taken from the soil vapor monitoring probes and system exhaust sampling ports are used to monitor the progress of cleanup, to estimate the volume of hydrocarbons removed by the system, and to establish a timetable for completion of the project.

Where appropriate, ground water samples will be taken periodically for analysis by EPA 600 series tests; this will allow an evaluation of the remedial effect of vapor extraction technology on the ground water contamination due to accelerated intraphase transfer from the ground water to the overlying soil gas.

System Shut Down and Demobilization

When monitoring indicates that remediation goals have been achieved, the system will be shut down. The pump and all aboveground piping will be removed from the site. All vapor probes, extraction wells, and below grade manifold piping are typically left in place. If necessary, confirmatory soil samples may be taken, and an ongoing analysis of soil vapors at vapor probe locations can be undertaken to ensure cleanup.

COST COMPARISON

The remediation of contaminated soils by vapor extraction/soil venting in the past few years has demonstrated the potential effectiveness and economics of this technology. Generally, vapor extraction can be applied for as little as $10 to $50 per cubic yard. This compares quite favorably to other treatment and disposal alternatives. For example, excavation and landfill or incineration disposal can cost as much as $200 - 350/yd^3.

REFERENCES

1. Porter, J. Winston, 1989. "Superfund Progress": Hazardous Material Control, Volume 2, No. 1, Page 48.

2. Baehr, A.L., 1987, "Selective Transport of Hydrocarbons in the Unsaturated Zone Due to Aqueous and Vapor Phase Partitioning": Water Resources Research, Vol. 23, No. 10, Page 1926-1938.

3. Sleep, B.E., and Sykes, J.F., 1989, "Modelling the Transport of Volatile Organics in Variably Saturated Media": Water Resources Research, Vol. 25, No. 1, Page 81-92.

4. Thornton, J.S., and Wooton, W.L., 1982, "Venting for the Removal of Hydrocarbon Vapors from Gasoline Contaminated Soil", Journal of Environmental Science and Health, A 17, Page 31-44

5. Marley, M.C., and Hoag, G.E., 1984, "Induced Soil Venting for Recovery/restoration of Gasoline Hydrocarbons in the Vadose Zone"; Proceedings of the National Water Well Association American Petroleum Institute Conference on Petroleum Hydrocarbons and Organic Chemicals in Groundwater, Nov. 5-7, Houston, TX.

6. Baehr, A.L. Hoag, G.E., and Marley, M.C., 1989 "Removal of Volatile Contaminants from the Unsaturated Zone by Inducing Advective Air Phase Transport": Journal of Contaminant Hydrology, Vol. 4 Feb., Pages 1-26.

7. Krishnayya, A.V., O'Connor, M.J., Agar, J.G., and King, R.D., 1988 "Vapour Extraction Systems - Factors Affecting their Design and Performance,": Proceedings of the National Water Well Association - American Petroleum Institute Conference on Petroleum Hydrocarbons and Organic Chemicals in Groundwater, Nov., Houston, Texas, Pages 547-569

8. Regalbuto, D.P., Barrera, J.A., and Lisieki, J.B. 1988, *In-situ* Removal of VOC's By Means of Enhanced Volatilization": Proceedings of the National Water Well Association - American Petroleum Institute Conference on Petroleum Hydrocarbons and Organic Chemicals in Groundwater, Nov., Houston, Texas, Pages 571-591.

9. Johnson, P.C., Kemblowski, M.W., and Colthart, J.D., 1988, "Practical Screening Models for Soil Venting Applications": Proceedings of the National Well Water Association - American Petroleum Institute Conference on Petroleum Hydrocarbons and Organic Chemicals in Groundwater, Nov., Houston, Texas, Pages 521-547.

10. Towers, D., Dent, M.J., and Van Arnam, D.G., 1989, "Choosing a Treatment for VHO-Contaminated Soil", Hazardous Material Control, Vol. 2, No. 2,

Page 8.

11. Marley, M. C., 1985, Quantitative and Qualitative Analysis of Gasoline
    Fractions Stripped by Air from the Unsaturated Zone"; M.S. Thesis,
    University of Connecticut, Department of Civil Engineering, Page 87.

12. Bruell, C.J., and Hoag, G. E., 1984, Capillary and Packed Column Gas
    Chromatography of Gasoline Hydrocarbons and EDB. Proc. National Water
    Well Association/American Petroleum Institute Conference on Petroleum
    Hydrocarbons and Organic Chemicals in Groundwater, Nov. 87, Houston, TX,
    Pages 234-266.

13. Cliff, B.L. 1988, "Method Development for the Gas Phase Analysis of
    Gasoline and Solvent Contaminated Soil," M.S. Thesis, University of
    Connecticut, Department of Civil Engineering.

14. Hoag, G.E., and Marley, M.C. 1986 "Gasoline Residual Saturation in
    Unsaturated Uniform Aquifer Materials." ASCE, Environmental Eng.
    Division, Vol. 112, No. 3, Pages 586-604.

15. Bruell, C.J., 1987, The Diffusion of Gasoline - Range Hydrocarbon Vapors
    in Porous Media. Ph.D. Dissertation, University of Connecticut, Civil
    Engineering Department, 157 pages.

16. Muskat, M., and Botset, H.G., 1931,1. "Flow of Gas through Porous
    Materials": Physics, Vol. 1, Pages 27-47.

17. Boardman, C.R., and Skrove, J.W., 1966, "Distribution in Fracture
    Permeability of a Granitic Rock Mass Following a Contained Nuclear
    Explosion": Journal of Petroleum Technology, Vol. 181, No. 5, Pages
    619-623.

18. Stallman, R.W., and Weeks, E.P., 1969, "The Use of Atmospherically Induced
    Gas Pressure Fluctuations for Computing Hydraulic Conductivity of the
    Unsaturated Zone": Geological Society of American Abstracts with
    Programs, Pt. 7, Page 213.

19. Weeks, E.P, 1977, Field Determination of Vertical Permeability to Air in
    the Unsaturated Zone: U.S. Geological Survey Open-file report 77-346 92
    p.

20. Rosza, R.B., Snoeberger, D.F., and Bauer, J., 1975, Permeability of a
    Nuclear Chimney and Surface Alluvium: Livermore Lab Report UCID-16722,
    11 p.

21. Kirkham, D., 1946, "Field Methods for Determination of Air Permeability of
    Soil in its Undisturbed State": Soil Science Society of America
    Proceedings, Volume 11, Page 93-99.

22. Evans, D.D., and Kirkham, D., 1949, "Measurement of Air Permeability of
    Soil *In-situ*": Soil Science Society of America Proceedings, Volume 14,
    Page 65-73.

23. Grover, B.L., 1955, "Simplified Air Permeabilities for Soil in Place":
    Soil Science Society of America Proceedings, Volume 19, Pages 414-418.

24.Tanner, C.B., and Wengel, R.W., 1957, "An Air Permeameter for Field and
    Laboratory Use":   Soil Science Society of America Proceedings, Volume
    21, Pages 663-664.

25. Klinkenberg, L.J., 1941, The Permeability of Porous Media to Liquids and
    Gases: American Petroleum Institute Drilling and Production Practice.

26. Baehr, A.L., and Hult, M.F. 1988, "Determination of Air-phase Permeability
    at the Bemidji Research Site": USGS 4th Toxic Substances Hydrology
    Technical Meeting, Sept., Phoenix, Arizona.

27. Brooks, R.H., and Corey, A.T. 1966, "Properties of Porous Media Affecting
    Fluid Flow": Journal of the Irrigation and Drainage Division,
    Proceedings of A.S.C.E., June Pages 61-88.

28. Parker, J.C., Lenhard, R.J., and Kuppusamy, T. 1987, "A Parametric Model
    for Constitutive Properties Governing Multiphase Flow in Porous Media":
    Water Resources Research, Vol. 23, No. 4, Pages 618-624.

Appendix G
Modeling Applications to Vapor Extraction Systems

Lyle R. Silka[a], Hasan D. Cirpili[a], and David L. Jordan[a]

INTRODUCTION

Vapor extraction systems (VESs) can be an effective tool in remediating unsaturated soils contaminated by volatile organic compounds (VOCs). The operating principles are straightforward. Contaminated soil is flushed with fresh air via a vacuum extraction well, drawing VOCs from the soil. As the contaminant is drawn off, more VOCs go into the vapor phase to regain equilibrium, and are again drawn off by the vacuum. In a clean, homogeneous material this procedure should work quite well to remove the VOCs.

However, there are potentially severe limitiations for a VES operation. Heterogeneous soil may contain isolated or dead-end pores and have variable saturation, and thus not be strongly affected by the VES. Since advective movement of the VOC in these stagnant areas is negligible, diffusion of the contaminant will dominate. Diffusive movement tends to be much slower than advective movement, and thus would impede the efficiency of a VES.

In a soil where some regions allow transport of VOCs by advection and other regions limit VOC transport to diffusion, the advection dominated portions of the soil would be quickly flushed of VOCs, while flushing the diffusion dominated portions would take much longer. The VOC extraction rate from such a soil would be high in the early stages of operation, during the advection dominated portion of flushing. Once the advection dominated region had been flushed out, the extraction rate would tend to exhibit considerable tailing, as VOCs slowly diffused out of the stagnant regions of the soil and into the advection dominated region. Therefore, the extraction rate of a VES in the early stages of operation should not be extrapolated to the entire flushing period, since it may only reflect the rate of contaminant removal from the advection dominated area of the soil. Thus, there is a distinct need to quantify the VES process, and especially the effects of a diffusion dominated region.

---

[a]HYDROSYSTEMS, P.O. Box 348, Dunn Loring, VA 22027

PREVIOUS STUDIES

Previous studies include a one-dimensional model developed by Baehr and Hoag, (1987) which describes advection and diffusion of gasoline vapors in an unsaturated sand column. They found that the rate of volatilization of organics was high relative to the rate of air flushing in areas of fairly open porosity. However, they also mention that this may not hold true in areas where vapor extraction is diffusion limited. Their results also show that even in an advection dominated system, the extraction rate will exhibit exponential decay, due to a shift in the organic phase to less volatile components, as the more volatile components are flushed out first.

Bowman (1987) cites empirical results from a VES installed at the Twin Cities Army Ammunition plant in Minnesota, which displays an exponential decline in extraction rate (Figure G-1). From the figure, it is apparent that the extraction rate is quite high in the early stages of operation, but drops off quickly and approaches a steady state value asymptotically. Even after adding another blower at 120 days, the rate of extraction remained essentially constant at the asymptotically low value throughout the remainder of operation. The flattening of the curve is probably due to both fractionation of VOCs and diffusion dominated transport.

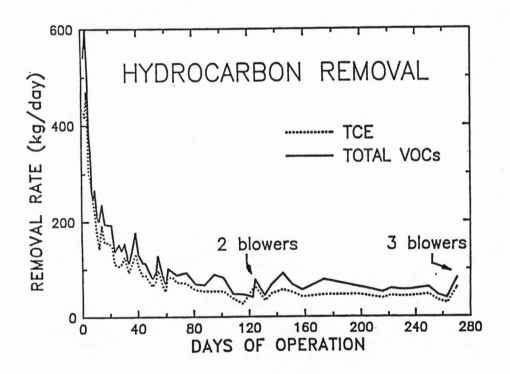

Figure G-1. Hydrocarbon Removal from the Twin Cities site.

Woodward and Clyde (1985) also present empirical data from a contaminated site in Tacoma, Washington (Figure G-2). Their data indicate that the extraction rate reached an equilibrium value. One possible explanation of their data is that the rate of air flushing is lower than the rate of volatilization, thus the vapor being extracted was continually being re-saturated with VOCs. Therefore, the system remained in the advection dominated stage, and did not reach the stage where the rate of VOC removal would be limited by diffusion. The Lauck's Laboratory dichloroetheylene (DCE) curve does show that the rate of extraction drops off significantly after an initial plateau, and remains fairly constant for three days. This may represent a transition from the advection dominated to the diffusion dominated regime.

CONCEPTUAL MODEL

The proposed conceptual model consists of a three layer case, shown in Figure G-3. A high porosity, low soil moisture middle layer (Region 1) is bounded by two low porosity, saturated layers (Region 2). We assume that a vacuum applied at the right hand boundary creates a constant flux of soil gas out of the system. Also, we assume that contaminant transport in Region 1 is advection dominated, and that contaminant transport in Region 2 is diffusion dominated. VOCs in the soil gas in Region 1 should be flushed from the system fairly rapidly since the water content in Region 1 is low and the soil porosity is high. As Region 1 is flushed and a concentration gradient is established from Region 2 into Region 1, VOCs will begin to diffuse from Region 2 into Region 1, volatilize into the soil gas, and be advectively flushed from the system. We used trichloroethylene (TCE) as our generic VOC in this study.

For the purposes of this modeling, a simple one-dimensional mass balance approach is used. An initial contaminant concentration is given for the Region 2. The concentration of VOC in the soil gas in Region 1 is then given by Henry's law

$$C_G = K_H C_L$$

where C_G is the concentration in the soil gas in Region 1, C_L is the concentration in the liquid in Region 2, and K_H is Henry's constant.

The change in mass of contaminant in each of the two regions is acounted for by a simple finite-difference equation. In Region 1, both diffusion and advection are considered, although advection will be the dominant transport process. Using a simple finite-difference formulation, the change in VOC concentration in Region 1 is:

$$dC_1/dt = D_E dC_1/dz + vC_1$$

where C_1 is the concentration in Region 1, D_E is the effective diffusion coefficient, v is the velocity of the soil gas, and z is the vertical direction. Silka (1988) derives D_E, which is an effective diffusion coefficient for the VOC which considers partitioning, adsorption, and tortuosity.

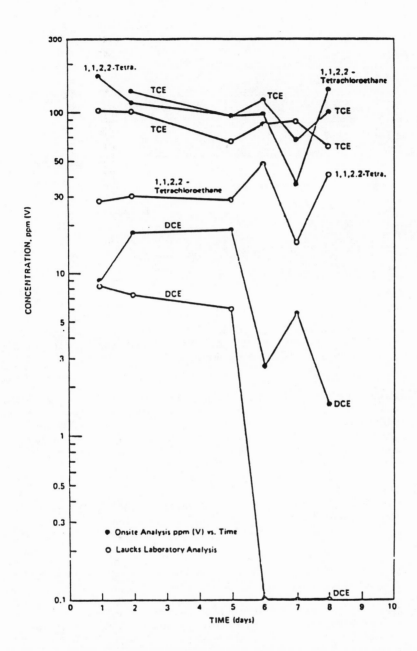

Figure G-2.   Concentration of Extracted Soil Gas from the Tacoma Site.
(After Woodward and Clyde, 1985).

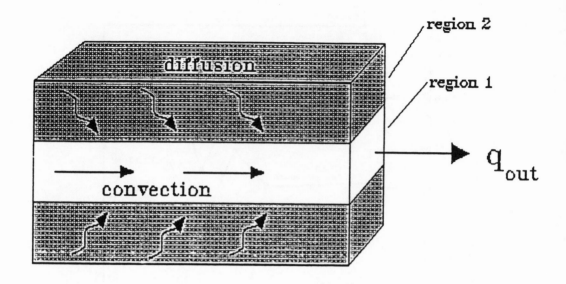

Figure G-3. Conceptual Model Used for the Numerical Modeling
          The dark regions are diffusion dominated, while the light
          region is convection dominated

Diffusion from the liquid layer in Region 2 to the gas layer in Region 1 is assumed to be controlled by the concentration gradient between the two layers. The diffusion coefficient is given by a turbulent diffusion coefficient (Thibodeaux, 1979) for diffusion from a liquid into a gas. Thus the mass balance equation for the liquid layer in Region 2 is:

$$dC_2/dt - D_t dC_2/dz$$

where C_2 is the liquid concentration in the Region 2 layer, and D_t is the turbulent diffusion coefficient.

<u>Assumptions</u>

The following assumtions were made in the course of developing this model and the associated modeling results. These are:

(1)   There is no volatilization loss to the atmosphere (i.e., the top of the system is an impermeable boundary). This could be realized in the field by sealing the ground surface with plastic or asphalt.

(2)   The adsorbed phase is neglected (i.e., it is assumed that VOCs will not desorb back into the system from organic matter or soil solids). There is no partitioning between soil solids and soil gas or water.

(3)   Diffusion in the Region 2 layer is Fickian.

(4)   All soil properties are homogeneous in each soil region.

(5)   VOC soil-gas concentrations achieve equilibrium instantaneously (i.e., the rate of extraction is significantly slower than the rate of volatilization of VOCs).

(6)   Model geometry is linear and one-dimensional.

Results

Results of the numerical model are shown in Figures G-4a and G-4b. Reasonable values of extraction rate, q, were chosen and soil gas velocities were calculated from these. The extraction values chosen seemed small (10^{-4} - 10^{-3} cfm), compared with values used at several actual sites (typically 10^1 - 10^2 cfm). However, much of the flux in the actual field tests may be due to short-circuiting with the atmosphere, and thus may not be representative of advection through the porous media. In our model, we assume that all of the gas flux in the system acts to extract the soil gas.

Note the general character of the concentration curves, which show an initial region of rapid extraction, followed by a steep decline in the rate of extraction. The total concentration in the extracted soil gas approaches an asymptotic value, which is controlled by diffusion in the liquid layers. The initial character of the concentration curve is controlled mainly by the extraction rate, q. A low extraction rate (depicted in Figure G-4) obviously takes longer to flush out the initial soil gas in Region 1.

Both curves compare favorably with results cited in Bowman (1987), shown in Figure G-1. The model results have the same initially high rate of extraction, followed by a rapid exponential decline in the rate of extraction. However, the model only compares favorably with one curve from the Woodward and Clyde data (1985), the DCE data from Lauck's Laboratory. This curve shows an initial high concentration in the system, followed by a rapid decrease in concentration, dropping off to a residual value of about 0.1 ppm. The other Woodward and Clyde curves do not compare as favorably to model predictions. This may be due the much higher volatility of DCE and lower mass of DCE in the soil, or the possibility that the TCE curves depict situations where TCE remains in the region where the high porosity zone is still being flushed out, and rate limiting diffusion has not come into effect.

CONCLUSIONS

Numerical modeling results of a vapor extraction system qualitatively agree with results from several empirical studies. The proposed model consists of two zones: an unsaturated, high porosity region, which is advection dominated, and a saturated, low porosity zone, which is diffusion dominated. The response of such a system to a VES has been shown to consist of an initial period of rapid vapor extraction, whereby the advective zone is flushed of soil gas. The advective period is followed by a sharp exponential decrease in the extraction rate as the system becomes diffusion dominated.

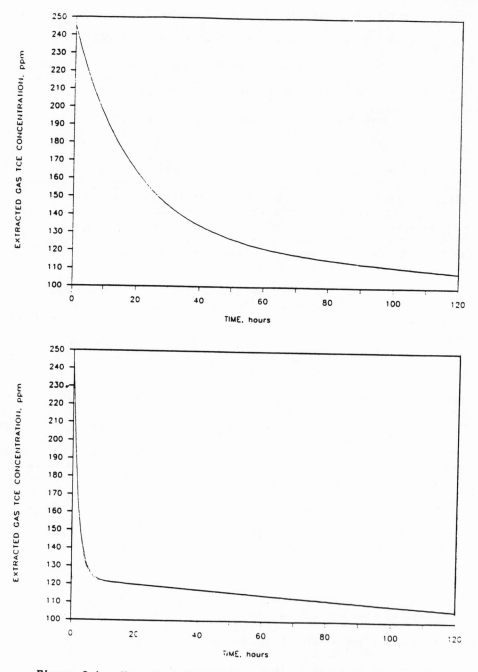

Figure G-4.   Numerical Modeling Results, (a) q = 0.0001 cfm,
(b) q = 0.001 cfm, initial concentration is 500
ppm TCE in the saturated zone.

Thus, it is important to consider the diffusion stage when designing and pilot testing a VES, since predictions of extraction time for a given remedial action based on the early response of the VES can be erroneous.

One other factor should also be considered in designing a VES. In a highly adsorptive soil, such as a soil with a high organic content, a large amount of contaminant would be resident in the solid matrix of the soil. This contaminant could easily be desorbed and released to the water and/or gas phases in the soil, and thus re-contaminate the soil gas. This would necessitate further soil gas extraction. The amount of contamination in the different phases of the soil could easily be measured to gauge the magnitude of these effects. Effects of this situation are currently being investigated.

Appendix H
Performance of In Situ Soil Venting Systems
at Jet Fuel Spill Sites

David W. DePaoli[a], Stephen E. Herbes[a], and
Michael G. Elliott, Capt, USAF, BSC[b]

INTRODUCTION

The Air Force Engineering and Services Center and Oak Ridge National
Laboratory (ORNL) are performing a field demonstration of in situ soil venting
at a 27,000-gal jet fuel spill site at Hill Air Force Base (AFB), Utah. In
situ soil venting is a soil cleanup technique that uses vacuum blowers to pull
large volumes of air through contaminated soil. The air flow sweeps out the
soil gas, disrupting the equilibrium existing between the contaminants on the
soil and in the vapor. This causes volatilization of the contaminant and
subsequent removal in the air stream. In situ soil venting has been used for
removing volatile contaminants such as gasoline and trichloroethylene, but a
full-scale demonstration for removing jet fuel from soil has not been
reported.

This paper briefly describes the jet fuel spill site and the design and
results to date of our full-scale in situ soil venting system.

DESCRIPTION OF JET FUEL SPILL SITE

On January 9, 1985, in a fuel yard at Hill AFB, Utah, ~27,000 gal
(102,000 L) of jet fuel (JP-4) spilled on the ground after an automatic
filling system malfunctioned and underground storage tanks overfilled. JP-4
is made by blending various proportions of distillate stocks such as naphtha,
gasoline, and kerosene to meet military and commercial specifications. In
general, it has more heavy molecular weight hydrocarbons and is less volatile
than gasoline and other contaminants which have previously been investigated
for remediation by in situ soil venting.

An initial surface cleanup effort at the spill site resulted in the
recovery of about 1000 gal (3800 L) of JP-4 with the remaining portion

[a]Engineering Development Section, Chemical Technology Division, Oak Ridge
National Laboratory*, Oak Ridge, Tennessee 37831-6044
[b]Air Force Engineering and Services Center, United States Air Force, Tyndall
Air Force Base, Florida 32403-5001
*Operated by Martin Marietta Energy Systems, Inc., for the U.S. Department of
Energy under contract DE-AC05-840R21400

infiltrating into the soil. A soil sampling study conducted in December 1985 delineated the areas of soil having >1% fuel in the soil. Based on this study, a decision was made to excavate the highly contaminated soil near the underground storage tanks and then place the tanks in an aboveground concrete enclosure.

To determine if the JP-4 had the potential to reach the water table, further investigations were completed to evaluate the geohydrological characteristics and the contaminant level in the soil of the site. Investigations included seismic and resistivity tests, soil vapor surveys, and core-boring analysis.

Based on past geophysical investigations at Hill AFB, it was known that the Provo formation comprises the surface strata beneath the spill site. The Provo formation consists of medium to fine sands with thin, interbedded layers of silty clay. Regionally, these sands are underlain by clay layers that extend to a depth of 600 ft (180 m) below land surface (BLS) at a well which is located 500 ft (150 m) south of the spill site.

A total of 43 soil borings were performed for characterization of the spill site. The lithologic logs describe a surface layer of brown silty sand about 4 ft (1.2 m) thick, underlain by brown sand to a depth of 23 to 35 ft (7 to 10.7 m) throughout the spill area. Variable-spaced clay layers were reported at depths between 23 and 42 ft (7 and 12.8 m).

The Delta aquifer, at an average depth beneath Hill AFB of ~600 ft (180 m), is the regional aquifer of greatest significance as a water-bearing unit because of its high permeability. The Sunset aquifer, at a mean depth of 300 ft (90 m) BLS is less permeable, and no wells of large volume draw from this unit. Both aquifers are isolated from the surface by impermeable formations which give rise to artesian flows in some wells in the area.

Local perched groundwater is found above the clay layers that confine the regional aquifer. Perched groundwater was encountered in one borehole near the fuel tanks at a depth of 32 ft (9.8 m) BLS. Perched water was also encountered at a depth of 51 ft (15.5 m) BLS, while water was present at a depth of 57 ft (17.4 m) BLS in a monitoring well. Saturated conditions were found at the clay layer [~27 ft (8.2 m) BLS] in several boreholes beneath the excavated tanks. The seismic and resistivity tests indicated the presence of perched groundwater at a depth of 43 to 46 ft (13.1 to 14.0 m) BLS. However, data from soil borings suggest that the perched groundwater is variable in depth and probably not continuous throughout the spill site.

A soil gas survey was conducted at the spill site in 1986, with probes installed to a depth of 10 ft (3.0 m). Highest values extended from the point of fuel spillage west across the spill area, approximately along the path of fuel flow. A second survey was conducted in September 1987. Probes installed to a depth of 1 ft (0.3 m) within and outside the originally defined plume boundaries resulted in profiles of fuel vapor distribution in the soil that were virtually identical in areal extent to the earlier results. No fuel vapors were detected west of the fence that bounds the fuel storage area.

Analytical results from the soil boring samples showed the residual fuel concentrations to be highest between the surface and a depth of 10 ft (3.0 m) in the area of the spill. Little lateral movement appears to have occurred since the spill. Residues and fuel odors have been detected in two boreholes in the western portion of the spill area, indicating that downward migration of vapors and/or fuel to 50 ft (15.2 m) BLS has occurred in at least several locations. In the eastern portion of the spill area, fuel appears to have migrated to the surface of a clay layer [33 ft (10 m) BLS]. Beneath the tank excavation, fuel levels are highest between 18 and 23 ft (5.5 and 7.0 m) BLS and fuel was detected as deep as 33 ft (10.0 m) BLS.

Based on the site investigations, it was concluded that high levels of fuel hydrocarbons were present to depths less than 50 ft (15.2 m). Also, because no continuous confining layer was identified at the spill site, a possibility exists for downward migration of the JP-4. Therefore, a no-action alternative was not applicable at the site, and a remediation technique must be implemented to prevent the contamination of groundwater. Since the soil is very sandy and high permeabilities for air flow were expected, the Hill AFB spill site provided an ideal setting for investigation of in situ soil venting as a remediation technique for the JP-4 spills.

DESIGN OF FULL-SCALE IN SITU VENTING SYSTEM

Based on information from the site characterization and a one-vent pilot test that was conducted in January 1988,[1] a full-scale in situ soil venting system was designed for collection of data and remediation of the JP-4 contaminated soil. The venting system was designed to consist of the three subsystems (Figure H-1): (1) a vertical vent array in the area of the spill,(2) a lateral vent system under the new concrete pad and dike for the tanks, and (3) a lateral vent system in the pile of soil that was retained after excavation for the tanks. This design includes features that permit evaluation of several factors affecting contaminant transport and subsurface air flow.

The vertical vent subsystem consists of 16 vents and 31 pressure monitoring points, covering an area of 120 x 100 ft (36.6 x 30.5 m). Half of the vertical venting area is covered by a plastic liner for comparison of flow patterns with and without a surface barrier. The vertical vents were located based on the best knowledge of the contaminant distribution, allowing flexibility in operation to investigate different venting strategies. The vents are arranged in a square grid with a 40-ft (12.2-m) spacing. The center line of vents has a 20-ft (6.1-m) spacing and is aligned from the existing vent installed for the single-vent pilot test to the point at the tank where the spill occurred. The 20-ft (6.1-m) spacing is not intended to be an optimized vent spacing; rather, it is used to allow operation of several vent configurations. Each vent is valved separately to allow each to act as either an extraction vent or as a passive inlet vent. The vents were constructed of 4-in.-(10.2-cm-)ID schedule 40 PVC screen [slot width of 0.02 in. (.05 cm)] and were installed in a 9 5/8-in. (24.4-cm) augered hole. Flush-joint schedule 40 PVC was used for riser pipes. They were screened between 10 and 50 ft (3.0 and 15.2 m) BLS and capped at the lower end.

CONCEPTUAL DRAWING OF IN SITU SOIL VENTING DEMONSTRATION SYSTEM HILL AFB

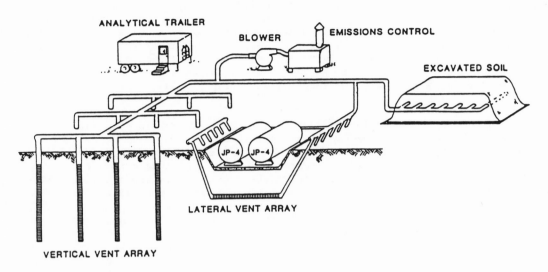

Figure H-1.   Conceptual Drawing of In Situ Soil Venting Demonstration System Hill AFB

The pressure-monitoring points for the vertical vent subsystem were located to map the pressure distribution at various depths in order to determine air flow patterns resulting from different venting strategies. The pressure monitoring points may be broken down into three groups: (1) points surrounding the vents in the area with a surface barrier, (2) points surrounding the vents in the area without a surface barrier, and (3) points surrounding the entire spill system to determine areal vacuum influence. The configuration of the pressure monitoring points is such to provide pressure distribution data while minimizing the influence of soil inhomogeneities. Each pressure monitoring point is installed in a separate borehole to avoid uncertainty invited by boreholes with multiple completions. The pressure-monitoring wells were constructed of 1-in.-(2.54-cm-)OD flush-joint PVC with a 2-ft (0.6-m) screened section, capped at the lower end.

Large diameter PVC pipe [6 to 8 in. (15.2 to 20.3 cm)] and fittings were installed to direct flow from the vents to the blower. Pipes were run on stands along the ground, meeting a manifold at the center of the vertical system. The pipes throughout the system were heat traced and insulated with fiberglass insulation, wrapped with aluminum covering.

The lateral vent subsystem under the new concrete pad was installed at the time of the tank excavation. This system is being used to investigate behavior of vents in horizontal geometry while decontaminating the soil beneath the tanks that was not removed during the tank excavation. The subsystem consists of six lateral vents constructed from 35 ft (10.7 m) of 4-in. (10.2-cm) perforated polyethylene drainage pipes that lie at a depth of about 20 ft (6.1 m) BLS, and at a distance of about 15 ft (4.6 m) apart. The vents induce air flow in the soil between a plastic liner at a depth of 15 ft (4.6 m) and an underlying clay layer at a depth near 26 ft (7.9 m). Each vent pipe is valved separately to allow each to act either as an extraction vent or as a passive air-inlet vent. Thirty-two soil-gas monitoring probes were placed in the soil at the time of the tank excavation. The probes are used for monitoring of pressure and sampling of soil gas below the new concrete dike.

The venting system in the excavated soil pile consists of a series of lateral vents placed in the contaminated soil that was removed during the tank excavation. Approximately 52,000 ft^3 (1,500 m^3) of this soil (initially within the zone contaminated to greater than 1 wt % of hydrocarbons) was removed and formed into a noncompacted pile. The pile is approximately 160 ft (48.8 m) long with a nearly triangular cross section that is 43 ft (13.1 m) at the base and 12 ft (3.7 m) high. Eight vents were placed in the pile at a nominal level of 5 ft (1.5 m) high and 18 ft (5.5 m) apart. The length of the vents within the pile is about 36 ft (11 m). To prevent erosion of the pile due to wind and rain, the pile was dressed and covered with a geotextile matting. The matting is a woven wood fabric with netting on both sides which allows for air flow but prevents soil from escaping. The lateral vents for both the pile and under the new concrete pad were constructed from 4-in. (10.2-cm) perforated polyethylene drainage pipe wrapped in filter fabric.

A blower/emission control system was installed for inducing air flow to three subsystems and for treating emissions as necessary to meet regulatory requirements. The two rotary-lobe blowers provide the capability for extraction of up to 2000 ft^3/min (57 std m^3/min) of gas from the three subsystems. In order to protect against potential hazards presented by combustible gas mixtures, flame arrestors were installed at the inlet to each blower; the blowers are controlled by an automatic shutdown system based on a combustible gas detector. Two catalytic oxidation units are used for conversion of the jet fuel hydrocarbons to carbon dioxide and water before discharge into the atmosphere. The propane-fired units differ in the configuration of their catalyst beds, one having a fluidized-bed design and the other containing a fixed-catalyst bed. The units are being evaluated in terms of economics and reliability, as well as hydrocarbon destruction efficiency. A knock-out drum, flowmeters, and gas monitors were also included in the system.

RESULTS

Operation of the Hill AFB full-scale in situ soil venting system began in December 1988. As of June 15, 1989, ~70,000 lb (32,000 kg) of hydrocarbons have been extracted from the JP-4-contaminated soil in 42 million ft^3 (1.2 million std m^3) of gas. Nearly all of the hydrocarbons were removed from the vertical vent system area, since the lateral system has been operated sparingly to this point and the pile has retained little contamination.

Progress made in decontamination of the site may be seen by examination of Figures H-2 through H-4. These figures present contours of the depth-

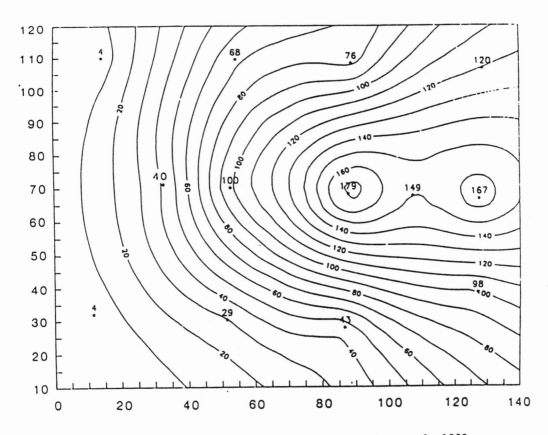

Figure H-2.   Soil Gas % LEL Contours - February 2, 1989.

averaged soil-gas concentration in the vertical vent area at two-month
intervals. The data were collected by extracting gas from each vent
separately and measuring the hydrocarbon levels in percent of the lower
explosion limit (LEL). The numbers on the plots refer to percent LEL, with
100 percent LEL equal to 13,100 ppm JP-4. The scale in each direction of the
plots is in feet, with the labelled asterisks denoting the position and
hydrocarbon concentration of each vent. As expected, the contours show a
trend of decreasing soil-gas hydrocarbon concentration, with the largest
changes occurring during the initial portion of venting.

Figure H-5 presents the hydrocarbon concentration measured in the
extracted gas as a function of the extracted gas volume. The data are
characteristic of soil venting operations, with initially high hydrocarbon
levels and a rapid decrease in concentration. Two discontinuities of interest
are present in the plot. The first is an abrupt decrease in concentration
from ~35,000 ppm hexane equivalent to 24,000 ppm, which occurred during a two-
week shutdown after about 42,000 ft^3 (1,200 m^3) of gas had been extracted.
This decrease is similar to a concentration decrease that was measured after

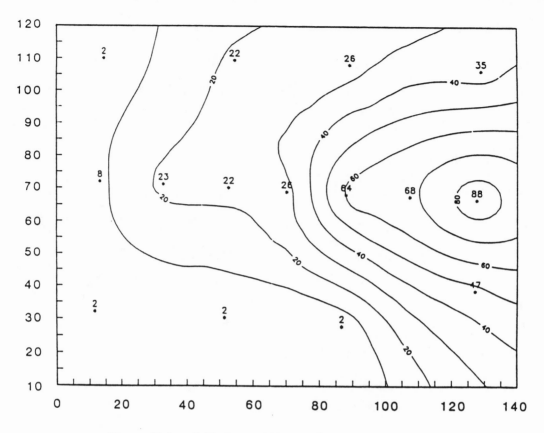

Figure H-3.   Soil Gas % LEL Contours - April 2, 1989.

~50,000 ft³ (1,400 m³) of gas was extracted during the pilot test at the site. It is postulated that this change in concentration is the result of the soil pore space in the zone of influence of the well being swept by the atmospheric air.   The equilibrated light fractions in the vapor are removed, and further hydrocarbon extraction is caused by volatilization.   The second discontinuity marks the point at which single-vent extraction was shifted from vent 7 (the third vent from the left in the center line of vents) to vent 10 (the second vent from the right in the center line).   The concentration increase is due to the fact that the latter well is centered in the most highly contaminated soil zone.   Other vents in the center line have since been included in operation with vent 10.

The concentration decrease shown in Figure H-5 deviates markedly from the straight-line logarithmic behavior that was projected for the system by noting the results of other researchers at single-component spills.[2-4]   This observed behavior may be explained in terms of the contaminant mixture present in the JP-4-contaminated soil.   Figure H-6 shows how the contaminant mixture affects the extracted gas hydrocarbon concentration by presenting the relative weights of different hydrocarbon fractions as measured in gas chromatographic

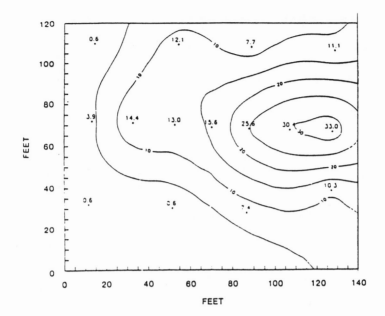

Figure H-4.   Soil Gas % LEL Contours - June 10, 1989.

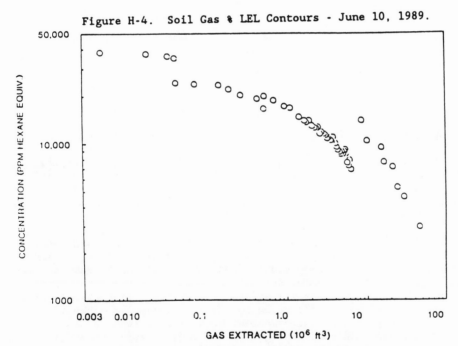

Figure H-5.   Hill Air Force Base Soil Venting
Extracted Gas Hydrocarbon Concentrations.

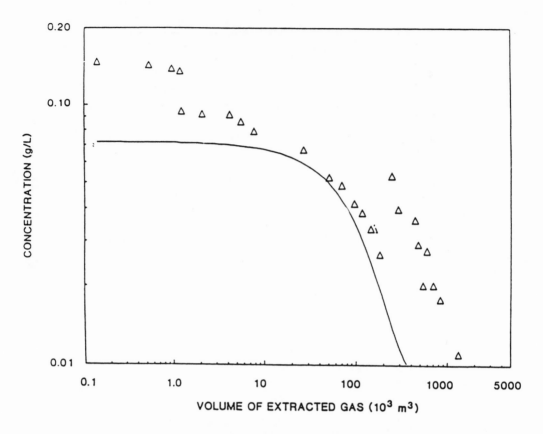

Figure H-6.   Comparison of Equilibrium Model and
Hydrocarbon Concentration Data.

analysis of gas samples. As was noted by Thornton et al.[5] in their pilot-
scale study of venting of gasoline from sand, the lighter hydrocarbon
fractions are a large portion of the hydrocarbons in the initial samples. As
the contaminant plume is depleted of the lighter fractions, the hydrocarbon
distribution in both the soil and the extracted gas is shifted in favor of the
heavier fractions, lowering the total concentration in the gas as a result of
the lower vapor pressures.

     A simple equilibrium model similar to that of Marley and Hoag[6,7] was
used for projection of system behavior. The model assumes perfect contact of
the gas with the hydrocarbons and equilibrium, as calculated by Raoult's law,
controlling transport to the vapor phase at each point in the soil. The
results obtained by input of a spill size of 26,000 gal (98,000 L) of jet fuel
with a distribution of components as measured in soil samples taken in October
1987 are shown in Figure H-7. Agreement with the actual results, in
particular in matching the shape of the curve, is quite good, given the
simplicity of the model and the uncertainties in the input parameters. Model
results are expected to improve with the input of hydrocarbon distributions
that were derived from soil samples taken during vent installation by a
sampling technique which was revised to prevent losses of lighter fractions.

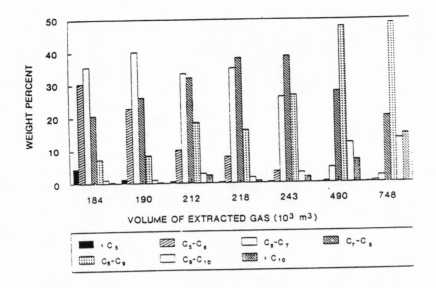

Figure H-7. Variation of Hydrocarbon Distribution in Extracted Gas.

It is expected that an equilibrium model may give good results in the earlier
that were derived from soil samples taken during vent installation by a
sampling technique which was revised to prevent losses of lighter fractions.
It is expected that an equilibrium model may give good results in the earlier
portion of venting, but it may deviate greatly from reality as soil
contamination is lowered, increasing the importance of (1) partitioning both
on soil and in the aqueous phase, and (2) airflow bypassing contaminated soil
zones.

A potentially significant means of hydrocarbon removal due to in situ
soil venting is enhanced biodegradation. The increased oxygen levels in the
soil gas due to infiltration of atmospheric air may considerably stimulate
biological activity. To evaluate this factor, carbon dioxide and oxygen are
being measured in the extracted gas. Initially, high CO_2 (11%) and low oxygen
(1%) levels were measured in the soil gas. As venting continued, the CO_2
levels decreased and oxygen levels increased, with an abrupt change in both
occurring at the same extracted gas volume noted for the change in hydrocarbon
concentration. Carbon dioxide levels have continued to be an order of
magnitude higher than background. From the start of operation until May 26,
approximately 40,000 lb (18,000 kg) of CO_2 had been extracted (after
subtraction of background CO_2 levels). If one considers the hydrocarbons as
the only carbon source (a reasonable assumption for the sandy soil), the CO_2
corresponds to about 12,700 lb (5,800 kg) of hydrocarbons. Considering that
half of the hydrocarbons consumed by bioactivity is converted to CO_2 and half
converted to biomass,[8] biodegradation may account for destruction of about
25,400 lb (11,500 kg) of hydrocarbons. This corresponds to 38% of the
hydrocarbons removed by volatilization. Thus, bioactivity contributed 27.5%
of total hydrocarbon removal.

A two-week shutdown of the system was performed in early June to evaluate biological activity. Gas samples taken from each vent after this period were analyzed for carbon dioxide and oxygen. The resulting contour plots are shown in Figures H-8 and H-9. Comparison of these plots with Figure H-4 shows that soil gas is depleted of oxygen and rich in CO_2 in areas where hydrocarbon levels are high, whereas a small change in oxygen from atmospheric levels and little or no CO_2 generation is seen in areas with low remaining hydrocarbon contamination. These results are important because they show that biological activity may continue to be significant despite no addition of nutrients. Biodegradation may prove to be the means for decontamination of heavier compounds that are not readily volatilized.

The results obtained thus far in the Hill AFB in situ soil venting demonstration have shown that this technique is very effective for removal of large amounts of jet fuel from soil in a very short period of time. It remains to be seen if the technique may achieve complete remediation of a jet fuel spill site. Our continued testing is aimed at answering this question as well as determining the importance of various factors in hydrocarbon removal. We will continue to sample the extracted gas to determine both the total hydrocarbon levels and hydrocarbon distribution. The effects of moisture on volatilization and bioactivity will be determined by monitoring soil moisture and extracted gas humidity. The effect of heat addition to the soil for need volatilization will be tested by routing heated air from the catalytic oxidation units to vents acting as air inlets. Also, soil sampling will be conducted in October 1989 to determine the extent of hydrocarbon removal to that point. This upcoming data should provide insight into whether in situ soil venting is a viable remediation technique for jet fuel-contaminated soil and will be valuable for modeling and application of venting to other sites.

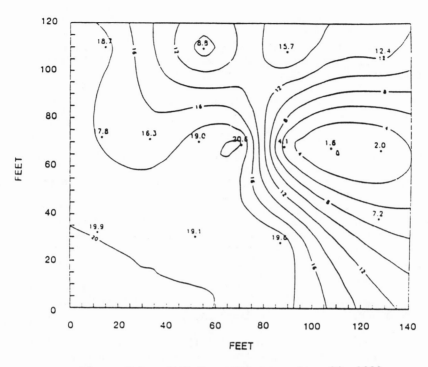

Figure H-8.   Soil Gas % Oxygen - June 10, 1989.

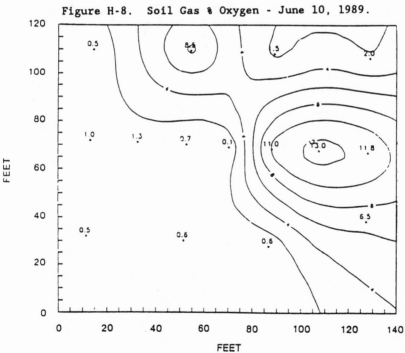

Figure H-9.   Soil Gas % Carbon Dioxide - June 10, 1989.

REFERENCES

1. Elliott, M. G. and D. W. DePaoli, "In Situ Venting of Jet Fuel-Contaminated Soil," presented at the 44th Purdue Industrial Waste Conference, May 10, 1989, West Lafayette, Ind.

2. Anastos, G. J., P. J. Marks, M. H. Corbin, and M. F. Coia, In Situ Air Stripping of Soils Pilot Study, Final Report, AMXTH-TE-TR-85026, October 1985.

3. Payne, F. C., C. P. Cubbage, G. L. Kilmer, and L. H. Fish, "In Situ Removal of Purgeable Organic Compounds from Vadose Zone Soils," In: Proceedings of the 41st Purdue University Industrial Waste Conference, West Lafayette, Ind., May 14, 1986. pp. 365–69.

4. Johnson, J. J., and R. J. Sterrett, "Analysis of In Situ Soil Air Stripping Data," In: Proceedings of the 5th National Conference on Hazardous Wastes and Hazardous Materials, HMCRI, Las Vegas, Nev., April 19–21, 1988.

5. Thornton, J. S., R. E. Montgomery, T. Voynick, and W. L. Wootan. "Removal of Gasoline Vapor from Aquifers by Forced Venting," 1984 Hazardous Material Spills Conference Proceedings, April 1984.

6. Marley, M. C. and G. E. Hoag, "Induced Soil Venting for Recovery/Restoration of Gasoline Hydrocarbons in the Vadose Zone," In: Proceedings of NWWA/API Conf. on Petroleum Hydrocarbon and Organic Chemicals in Groundwater - Prevention, Detection and Restoration, pp. 473–503, November 1984.

7. Marley, M. C. "Quantitative and Qualitative Analysis of Gasoline Fractions Stripped by Air from the Unsaturated Soil Zone," University of Connecticut, Master's Thesis, 1985.

8. Thornton, J. S. and W. L. Wootan, Jr., "Venting for the Removal of Hydrocarbon Vapors from Gasoline from Gasoline Contaminated Soil," J. Environ. Sci. Health A17(1), pp. 31–44, 1982.

Appendix I
In Situ Vapor Stripping Research Project:
A Progress Report

Robert D. Mutch, Jr, P.Hg.,P.E.[a], Ann N.Clarke, Ph.D.[a],
David J. Wilson, Ph.D.[b], and Paul D. Mutch[a]

A two-year long in situ vapor stripping research project is being
conducted at the Ciba-Geigy Plant in Toms River, New Jersey. The research is
being conducted by AWARE Incorporated and is co-funded by the USEPA Small
Business Innovative Research Program and Ciba-Geigy Corporation. The research
project, which began in August of 1988, involves a closely monitored
installation of in situ vapor stripping. The research program calls for the
vapor stripping facilities to operate for a period of one year. This paper
reports on findings through the first five months of operation.

The general objectives of the research program are to improve the
scientific foundation for this remedial technology, better define its
technical limitations and further refine the mathematical model of the
process, developed in an earlier phase of the research program (AWARE, 1987).
A further objective of the research is to evaluate the performance of
granulated activated carbon as a treatment agent for the extracted vapors.

The Ciba-Geigy Toms River Plant lies in the Atlantic Coastal Plain
Physiographic Province in Toms River, New Jersey. The site is underlain by
the Cohansey Sand, a geologic formation consisting predominantly of moderate
to high permeability sand, interbedded with finer-grained, often lenticular,
strata of silt and clay. The site chosen for the research program lies within
the central production area of the 1200 acre plant site at the location of
several recently demolished chemical process buildings. Soil contamination
was detected in the razing of the buildings, presumably from underground
storage tank and process pipeline leaks.

INITIAL SOIL CONTAMINATION LEVELS

A drilling program was undertaken in order to characterize initial
levels of soil contamination within the study area. The program consisted of
26 exploratory borings and collection and analysis of 40 soil samples. The
complete menu of organic priority pollutant analyses was run. Table I-1

---

[a]ECKENFELDER, Incorporated. 1200 MacArthur Boulevard, Mahwah, NJ 07430
[b]Senior Research Associate, ECKENFELDER, Incorporated; Professor of
  Chemistry, Vanderbilt University, Nashville, TN

summarizes the specific chemicals detected and their respective concentration
ranges (including limits of detection). Not all chemicals were identified as
present at all probe locations or at all depths at a given location. No acid
extractable compounds were detected at a limit of detection of 2.0 ppm (mg/kg)
with the exception of one sample which exhibited a phenol value of 3.0 ppm.
The most prevalent soil contaminants in the study area are
1,1-dichlorobenzene, 1,3-dichlorobenzene, 1,4-dichlorobenzene, and 1,2,4-
trichlorobenzene.

RESEARCH PROGRAM FACILITIES AND EQUIPMENT

    One of the initial tasks of the research proposal was the drilling of a
four-inch diameter extraction well in the approximate center of the area of
contamination. The well consisted of a four-inch diameter, five-foot long,
factory-slotted PVC screen which was set slightly above the water table on a
four inch diameter PVC casing. In the area of the project, the water table is
at a depth of approximately 20 feet. A series of 38 soil gas probes were
installed at radial distances of approximately 20, 40, 60, and 80 feet from
the extraction well. A number of the probes were constructed as clusters with
individual probes at depths of 5, 10, and 15 feet below ground surface. A
sketch of a typical soil gas probe is illustrated in Figure I-1. The probes
were constructed of Teflon tubing and were installed by means of a truck-
mounted hollow stem auger rig. The screened section of the probe was sand-
packed and the remaining annular space was sealed by bentonite pellets and
grout. The probes allow for measurement of in situ soil vacuum and also
permit sampling of soil gas quality. Twelve of the probes were fitted with
temperature thermisters to permit measurement of in situ soil temperature.

    Extraction of the soil gas vapors and much of the monitoring is
performed by AWARE's In Situ Vapor Stripping Pilot Unit. The eight-foot by
12-foot long pilot unit trailer houses two New York Blower Model 2606-A
pressure blowers. The blowers utilize a 26-inch aluminum compression fan
blade encased in a steel-frame housing. They are powered by 7-1/2 hp, 460
volt, 3-phase motor so at the rated 3500 rpm fan speed, the two blowers
produce 50.5 inches water column pressure on the outlet at a flow rate of 400
scfm. The blowers and associated ducts are configured for individual, series,
or parallel operation, depending upon flow rate/pressure requirements.

    The pilot unit trailer also contains a baffled demister to remove water
droplets from the air stream and instrumentation and controls for operation of
the system and the monitoring of system performance. A layout of the pilot
unit trailer is illustrated in Figure I-2. There are five sampling ports in
the duct work to allow sampling of extracted gas quality at various points in
the system. Measurements of temperature and pressure can also be remotely
taken at each sampling port. Treatment of the extracted gas is accomplished
by use of granulated activated carbon. A carbon canister is set up outside
the trailer as indicated in Figure I-2. An HNU Model PI-201 photoionization
monitor with an Esterline Angus Model 410 chart recorder is utilized to
continuously record gas quality. An electronic control panel, in conjunction
with a Masterflex pump, automatically samples each of the five gas monitoring
probes and a calibration gas cylinder once every hour. The automatic
sequencing can be overridden if manual readings are desired. The HNU
photoionization detector output data is stored on the chart recorder for

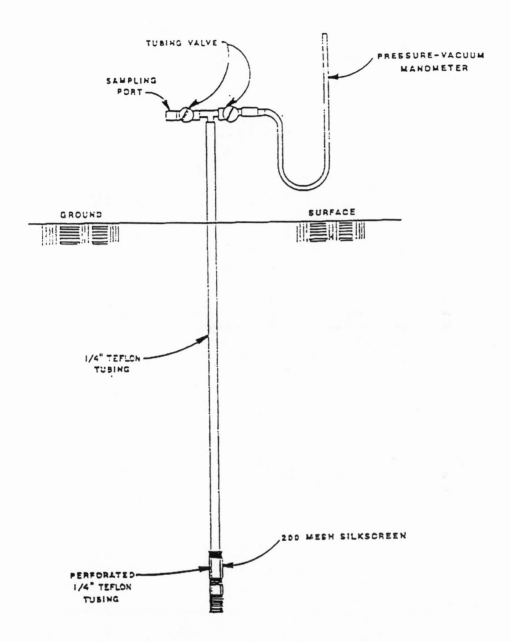

Figure I-1.   Sketch of Typical Soil Gas Probe.

DUCT SYSTEM OVERVIEW LAYOUT

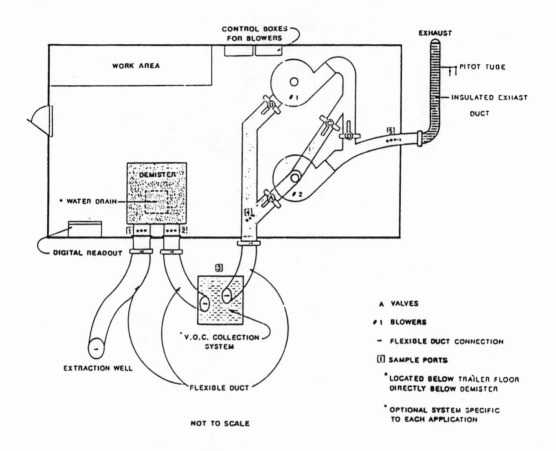

Figure I-2.   Layout of In Situ Vapor Stripping Pilot Scale
Research Trailer.

TABLE I-1.   Summary of Chemicals Identified at the Study Site.

Family	Chemical (limit of detection)	Range of Concentrations
Volatile	(ppb)	(ppb)
	Trichloroethylene (10)	ND-21
	1,2-dichlorobenzene (10)	ND-3,800,000
	1,1,2,2-tetrachloroethane (10)	ND-23
	Tetrachloroethylene (10)	ND-11
	1,1,1-trichloroethane (10)	ND-36.5
	Chloroethane (50)	ND-257
	Methylchloride (50)	ND-34
Base Neutral	(ppm)	(ppm)
	Benzidine (3)	ND-5.2
	Di-n-butylphthalate (1)	ND-3.4
	Fluoranthene (1)	ND-13
	Phenanthrene (1)	ND-15
	Pyrene (1)	ND-11
	1,2,4-trichlorobenzene (1)	ND-294
	Bis(2-ethylhexyl) phthalate (1)	ND-10
	Fluorene (1)	ND-1.4
	Indeno (1,2,3-cd) pyrene (1)	ND-2.2
	Anthracene (1)	ND-3.1
	Benzo(a) anthracene (1)	ND-5.3
	Benzo(a) pyrene (1)	ND-4.3
	Benzo(g,h,i) perylene (1)	ND-1.1
	Benzo(k) fluoroethene (1)	ND-4.2
	Chrysene (1)	ND-5.0
	1,3-dichlorobenzene (1)	ND-100
	1,4-dichlorobenzene (1)	ND-161
	Naphthalene (1)	ND-2.4
	Nitrobenzene (1)	ND-1.4

manual interpretation.   The flow rate of the system is monitored by means of a Dwyer pitot tube and micromanometer.

An air permit was obtained from the New Jersey Department of Environmental Protection in order to operate the vapor stripping system.   The permit established a maximum discharge concentration of 50 parts per million.

PRELIMINARY FINDINGS

The preliminary findings of the research project center upon the measured zone of influence of the extraction well, the quality of the extracted soil gas with time, the treatability of the extracted gas by means of the granulated activated carbon system, temperature variations occurring in the system, and the observed rise of the groundwater table induced by the vacuum extraction. The preliminary findings in each of these areas are briefly discussed as follows.

Zone of Influence

Mathematical modeling of the in situ vapor stripping process indicates that in an isotropic soil the zone of influence of an extraction well screened near the base of the unsaturated zone should produce a zone of influence with a radius approximately equal to the depth of the unsaturated zone (Wilson, et al, 1988). In a soil with vertical anisotropy, the radius of the zone of influence is increased in proportion to the degree of anisotropy. Because the Cohansey Sand was expected to have a vertical anisotropy of two to three, the in situ soil gas monitoring probes were set out at radial distances of 1D, 2D, 3D, and 4D, where "D" equals the depth to the water table area.

Soil gas extraction was commenced on September 6, 1988, at a rate of 180 cfm. In situ soil gas vacuum levels were observed almost immediately throughout the study area and reached a steady-state condition in less than 15 minutes. The in situ vacuum levels have remained essentially constant throughout the course of the research program. Contours of in situ vacuum levels are depicted in plan view and in cross-section in Figures I-3 and I-4,

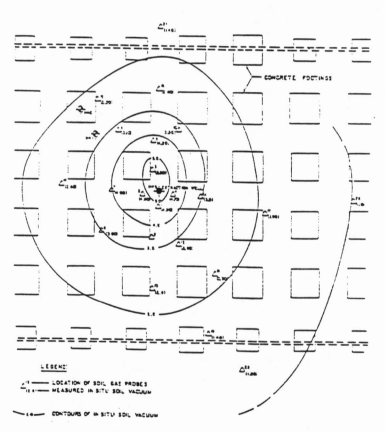

Figure I-3.   Contours of In Situ Soil Vacuum.

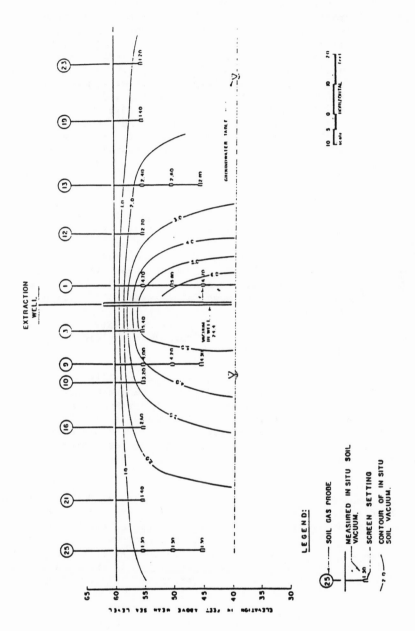

Figure I-4.   Cross-Sectional In Situ Soil Vacuum Contour Map
(inches of water).

respectively. As indicated in these figures, a wider zone of influence was established than anticipated, even considering the vertical anisotropy of the Cohansey Sand. It can be extrapolated from the measured in situ vacuum levels that the effective radius of influence of the soil gas extraction system is approximately 150 feet. This is more than twice the anticipated radius of influence. The apparent reasons for this occurrence can be tied to the surficial conditions. As mentioned earlier, the research project is sited at the location of previous production buildings. Approximately 25 percent of the surface of the study area is comprised of ten-foot square, five-foot thick, concrete footings. Moreover, the intervening soil around the footings is to a large extent fill material of a finer-grained character than the underlying native soils of the Cohansey Sand. Consequently, the upper five feet of soil acts to impede the influx of atmospheric air from the surface, causing the zone of influence to spread laterally beneath this surficial layer.

Extracted Soil Gas Quality

The quality of soil gas extracted during the initial months of the research program, measured by means of the photoionization detector, is presented in Figure I-5. The soil gas quality is presented with respect to days of system operation. Days of system shut down for maintenance and installation of additional granulated activated carbon canisters are omitted from the graph. Extracted soil gas concentrations were initially in the range of 110 to 140 parts per million and have fairly steadily declined to current levels of approximately 60 to 70 ppm. Chemical analysis of the extracted soil gas reveals that the principal constituents are: 1,1-dichloroethane; 1,1,2,2-tetrachloroethane; 1,1,1-trichloroethane; trichloroethylene; 1,2-dichlorobenzene; and 1,3-dichlorobenzene.

Figure I-6 presents a graph of discharge gas quality after granulated activated carbon treatment. The three peaks in the graph represent progressive exhaustion of the granulated activated carbon canisters. The first and third peaks on the graph represent exhaustion of the typical 1200 pound granulated activated carbon canisters used in the project. The intermediate peak represents a more rapid exhaustion of a standby granulated activated carbon system consisting of two parallel 55-gallon drum carbon canisters.

The treatment efficiency of the granulated activated carbon has been significantly diminished by sorption of water vapor in the carbon. The demister has removed relatively little water since the water occurs in the form of water vapor rather than as a mist. The demister is currently being replaced with refrigerated condenser unit, designed to remove approximately 75 percent of the water vapor from the gas stream. It is anticipated that this unit will prolong the useful life of the carbon by a factor of approximately four.

Temperature Variations

The temperature of the extracted gas, ambient air, and the 12 in situ probes have been measured throughout the course of the study. Figure I-7 depicts the variations in extracted soil gas temperature and ambient

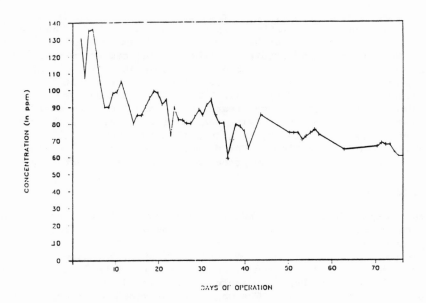

Figure I-5.   Extracted Gas Quality.

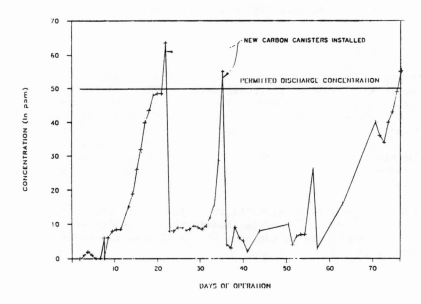

Figure I-6.   Discharge Gas Quality.

temperature. The temperature of the extracted soil gas was initially approximately 18°C and has steadily declined during the fall and beginning of winter to current temperatures of between 11°C and 12°C. Figure I-8 presents a graph of in situ soil temperature at probe 1. The temperature variations observed in probe 1 are characteristic of soil temperature variations occurring within the study area. The graph illustrates that, initially, soil gas temperatures were in the range of 18°C to 22°C. Also, the deeper soil probe exhibited a consistently lower temperature than the intermediate and shallow probes. This is not surprising considering the time of the year. With the onset of fall and winter, in situ soil temperatures declined and reversed their relative positions. The deeper probe, probe 1D, exhibited the highest temperature and the shallow probe, probe 1S, the coolest temperature.

Groundwater Quality

A rise in groundwater levels beneath in situ vapor stripping facilities has been both predicted and observed. The phenomenon results from the fact that the groundwater table represents the point in the subsurface where the voids in the soil or rock are not only fully saturated, but also at equilibrium with atmospheric pressure. Consequently, if soil gas pressures are reduced to below atmospheric pressure, a corresponding rise in the groundwater table should result. The magnitude of groundwater table rise (in inches) should coincide with magnitude of the pressure drop below atmospheric occurring at any point in the system (in inches). Monitoring of groundwater levels during the course of the research study confirms that the water table does indeed indeed rise a level commensurate with the soil vacuum levels produced by the extraction well. The maximum rise in the water table of nearly 2.5 feet occurred immediately beneath the extraction well.

MATHEMATICAL MODELING

A mathematical model has been developed for predicting various aspects of a full-scale in situ vapor stripping system. This model will be verified using data from the pilot scale study site discussed in this paper. The model was originally calibrated using laboratory data generated from specially designed equipment which simulated actual field parameters and operating conditions. A critical model parameter is the determination of lumped partition coefficients for the chemical constituents of interests. This coefficient addresses the chemical interaction with water, soil and other chemicals present.

The model can be used to generate important design criteria and optimize operating parameters from pilot scale studies for use in full-scale remediations. The model can be run on a PC. A list of the model capabilities is provided in Table I-2.

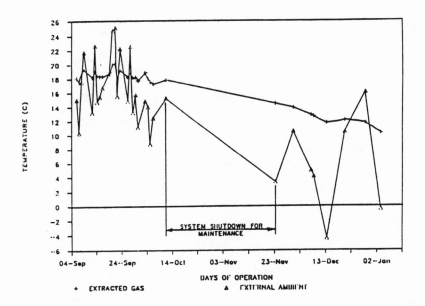

Figure I-7.   Extracted Gas and Ambient Temperatures.

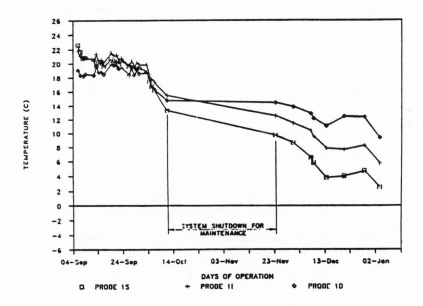

Figure I-8. Soil and Ambient Temperature.

Table I-2. List of Model Capabilities for the Prediction of
Design and Operating Parameters for Full Scale In Situ Vapor
Stripping.

---

1. Predict clean-up time to reach a target level of residual contamination.
2. Predict residual contaminant levels after a given period of operation.
3. Predict location of hot spots through diagrams of contaminant distribution.
4. Develop system design:
> horizontal well placement
> vertical well placement
> screen placement
5. Predict clean-up levels around buried debris from various system designs.
6. Predict impact of ambient air temperatures on removal.
7. Calculate the anisotropy of the soil or rock.
8. Predict recontamination time of the remediated vadose zone from slow moving contaminated groundwater.
9. Predict the rate of remediation of floating pools of LNAPLs.

FUTURE RESEARCH OBJECTIVES

While the preliminary findings have answered a number of questions concerning the behavior of in situ vapor stripping systems, the majority of the research objectives remain to be accomplished. These objectives include the following:

1. Describe the temporal variations in overall gas quality, as well as the relative proportions of individual constituents within the gas stream.

2. Determine residual levels of various contaminants in the soil at the conclusion of the project.

3. Describe the relationship between extracted gas flow and the resultant zone of influence and upon cleanup times.

4. Refine the mathematical model to permit modeling of more complex soil conditions such as layered heterogeneity.

5. Evaluate the performance of the refrigerated condenser unit upon gas stream humidity and granulated activated carbon life.

6. Determine the quality of condensate from the refrigerated unit.

REFERENCES

AWARE Incorporated, Phase I Zone 1 Soil Decontamination Through In Situ Vapor
    Stripping Processes, USEPA Contract No. 68-02-4446, April, 1987.

Wilson, D.J., A. N. Clarke, and J. H. Clarke. "Soil Clean-Up by In Situ
    Aeration, I. Mathematical Modeling," Separation Science and
    Technology, 23(10 & 11), pp 991-1037, 1988.

Appendix J
Soil Extraction Research Developments

George E. Hoag[a]

INTRODUCTION

Recently, in situ subsurface remediation processes have been the focus of significant attention by the scientific community involved with the clean-up of volatile and semi-volatile environmental contaminants. Of the in situ processes researched to date, vapor extraction holds perhaps the most widespread application to the remediation of these types of organic chemicals frequently found in the subsurface. The vapor extraction process has been successfully employed at many types of sites as a stand alone technology and may also be considered a synergistic technology to other types of in situ subsurface remediation technologies, such as, bioremediation and groundwater pump, skim and treat.

In the past five years, in situ vapor extraction has been applied at many sites by means of significantly different approaches. These range from "black box" design while you dig techniques to those utilizing sophisticated numerical models interfaced with laboratory, pilot and full-scale parameter determination for design purposes. The extent of success in field application of vapor extraction is varied, in many cases related to monitoring and interpretive limitations employed before, during and after the remediation. Because application of the technology is quite recent, many remediations are still in progress, thus final results interpretation and publishing in refereed scientific journals is limited.

RESEARCH MILESTONES IN VAPOR EXTRACTION

Thorton and Wootan (1982) introduced the concept of vertical vapor extraction and injection wells for the removal of gasoline product, as well as, vapor probe monitoring for the quantitative and qualitative analysis of diffused hydrocarbon vapors. A further enhancement of this research was published by Wootan and Voynick (1984), in which various venting geometries and subsequent air flow paths were hypothesized and tested in a pilot sized

---

[a]Director, Environmental Research Insitiute and Associate Professor of Civil Engineering, University of Connecticut, Storrs, CT 06269
[a]Senior Technical Consultant, VAPEX Environmental Technologies, 480 Neponset Street, Canton, MA 02021

soil tank. In their first study, 50 percent gasoline removal was achieved, while in their second study up to 84 percent removal of gasoline was observed.

Local Equilibrium Concept

In controlled laboratory soil column vapor extraction experiments by Marley and Hoag (1984) and Marley (1985), one hundred percent removal of gasoline at residual saturation was achieved for various soil types (0.225 mm to 2.189 mm average diameters), bulk densities (1.44 g/cm to 2.00 g/cm), moisture contents (0 percent to 10 percent v/v) and air flow rates (16.1 $cm^3/(cm^2$-min) to 112.5 $cm^3/(cm^2$-min.) They also successfully developed an equilibrium solvent-vapor model using Raoult's Law to predict concentrations of 52 components of gasoline in the vapor extracted exhaust of the soil columns.

Baehr and Hoag (1985) adapted a one dimensional three phase (immiscible solvent, aqueous and vapor phases) local equilibrium transport model developed by Baehr (1984) to include air flow as described by Darcy's Law for compressible flow. This first deterministic one dimensional model effectively predicted the laboratory vapor extraction results of Marley and Hoag (1984) and provided the basis for higher dimensional coupled air flow contaminant models for unsaturated zone vapor extraction.

Porous Media Air Flow Modeling

Because local equilibria prevailed in the above studies, a higher dimensional model, developed by Baehr, Hoag and Marley (1988) was used to model air flow fields under vapor extraction conditions. The three dimensional radially symmetric compressible air flow model is used to design vapor extraction systems using limited lab and/or field air flow pump tests. A steady state in situ pump test determination of air phase permeability is preferred over laboratory tests because an accounting is possible of the presence of an immiscible liquid, anisotropy, soil surface, variations in soil water conditions and heterogeneity in air phase permeabilities. The numerical solution developed can simulate flow to a partially screened well, and allow determinations of vertical and horizontal air phase permeabilities. Heterogeneous unsaturated zones can also be evaluated using the numerical simulation. Analytical solutions to radial flow equations, such as one developed early by Muskat and Botset (1931) are generally restricted to determination of average horizontal air permeability determination for impervious soil surfaces.

Removal of Capillary Zone Immiscible Contaminants

Hoag and Cliff (1985) reported that an in situ vapor extraction system was effective in removing 1330 L gasoline at residual saturation and in the capillary zone at a service station and achieved clean up levels to below 3 ppm (v/v) in soil vapor and below detection limits in soils. The entire remediation took less than 100 days. A groundwater elevation and product thickness log for the time period of before, during and after the vapor extraction remediation is graphically shown in Figure J-1. On day 250 only a skim of gasoline was present in this well and on day 290 no skim was detected. One year after the vapor extraction remediation took place, groundwater

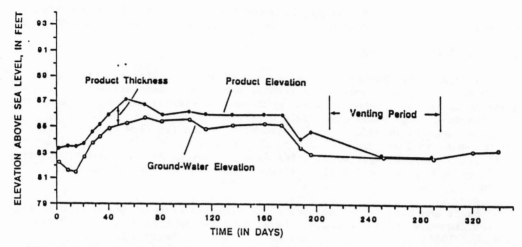

Figure J-1.   Apparent Product Thickness in Ground Water Monitoring Well.

samples were non detected for gasoline range hydrocarbons, reflecting that at least advective dispersive transport and possible natural microbiological activity in the groundwater were mechanisms responsible for this effect.

Field Application of Porous Media Air Flow Models

Baehr, Hoag and Marley (1988) utilized the above site for a field air pump test to determine the horizontal air phase permeability and to simulate the sensitivity of the model to changes in air phase permeabilities utilizing site geometries and boundary conditions. Based upon a full-scale air flow pump test, the air phase permeability for the site was predicted to be k - 7.0 x 10^{-8} cm^2 for a mass air withdrawal rate of 11.1 g/sec and a normalized pressure of P_s/P_{atm} - 0.9. For reference 11.1 g/sec, assuming an air phase density of 1.2 x 10^{-3} g/cm^3, equals about 555 L/min.

To illustrate the sensitivity of the model to a range of air phase permeabilities, the above service station vapor extraction well geometry, depth to water table, and appropriate boundary conditions were used as input and air phase permeability and mass air withdrawal rates were varied. In Figure J-2, the normalized air phase pressure in the well, for various mass air withdrawal rates and air phase permeabilities are shown. Significant increase in the vacuum developed in the wells can be observed for order of magnitude decreases in air phase permeabilities and small increases in the mass air withdrawal rates. Review of Figure J-2, indicates that if a mass air withdrawal rate of 40 g/sec was used at the service station site (k - 7.0 x 10^{-8}), the the normalized air phase pressure at the well would be approximately P_s/P_{atm} - 0.6, within an acceptable range of operating conditions. A limitation of the model developed by Baehr and Hoag (1988) is that it is not coupled to contaminant transport.   However, for the design of vapor extraction systems for volatile contaminants, this generally is not a fundamental need and can be accomplished by either laboratory venting tests, similar to those developed by Marley and Hoag (1984) or by utilization of the one dimensional coupled model developed by Baehr and Hoag (1985).

RESEARCH NEEDS

A fundamental need of vapor extraction modeling occurs in the area of capillary zone/unsaturated zone interaction when immiscible phases are present on or in the capillary fringe. While air phase modeling alone is probably adequate for most vapor extraction system system design purposes, particularly if a full three-dimensional model is used with optimization modeling, a rigorous modeling effort to couple air phase flow and immiscible contaminant transport, particularly in the capillary zone, will provide strategic insight to vapor extraction operation and planning.

To assess research needs in this area, two basic vapor extraction systems applications should be considered: 1. Immiscible contaminant with density less that 1.0 (petroleum range hydrocarbons); and 2. Immiscible contaminants with density greater than 1.0 (halogenated compounds).

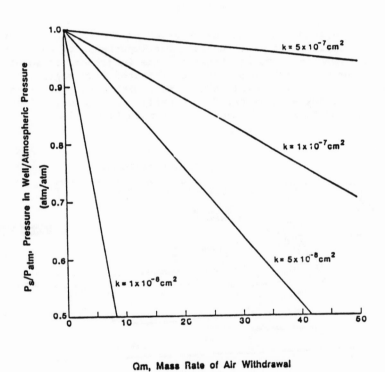

Figure J-2.   Normalized Air-Phase Pressure at a Single Well.

Generalized subsurface phase distributions for immiscible liquids with densities less than that of pure water are illustratively shown in Figure J-3. A typical vapor extraction system installation in this type of subsurface and contaminant condition is found in Figure J-4. In the case study presented by Hoag and Cliff (1985), as detailed above, pump and skim was employed at the site for the first 210 days of the remediation with only 300 L gasoline removed (i.e. mostly through manual bailing). Thus, an important question should be: Was vapor extraction alone necessary in this case or were both pump and skim and vapor extraction required for optimal or even effective remediation of immiscible contaminants? To answer this question requires an understanding of air-immiscible liquid-water three phase conduction and distribution in the porous media, particularly in the capillary fringe areas at a site. Additionally, the site history of groundwater fluctuation and immiscible contaminant behavior in the capillary fringe is essential information necessary to answer the above question. Parker, Lenhard and Kuppusamy (1986) and Lenhard and Parker (1986) provide a parametric model for three-phase conduction and measurements of saturation-pressure relationships for immiscible contaminants in the unsaturated and capillary zones. However to date, this author is not aware of the coupling of these types of models to air phase and contaminant transport models.

A more in depth hypothetical examination of the possible relationships near the capillary fringe will illustrate the importance of this zone in determining the need for pump and treat and the importance of solute mass transfer from the capillary zone into the saturated flow regime. In the case of a recent spill of an immiscible contaminant with density less than water, when relatively steady groundwater flow prevails, a zone may exists on the capillary fringe of floating product, as shown in Figure J-5.

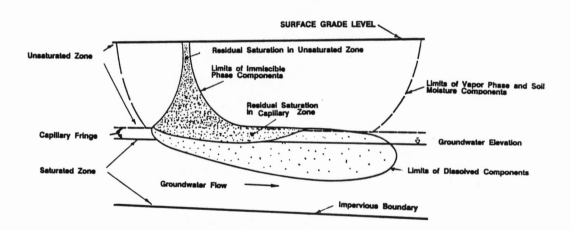

Figure J-3. Generalized Subsurface Phase distribution for Immiscible Contaminants with Densities Less than Pure Water.

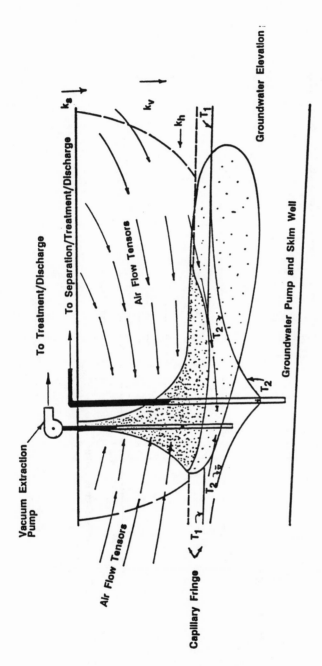

Figure J-4. Typical Vapor Extraction Pump and Treat In-Situ Subsurface
Remediation System for Immiscible Contaminants with
Densities Less than Pure Water. (T_1 represents the
groundwater table prior to pump and skim and T_2 is after
pump and skim).

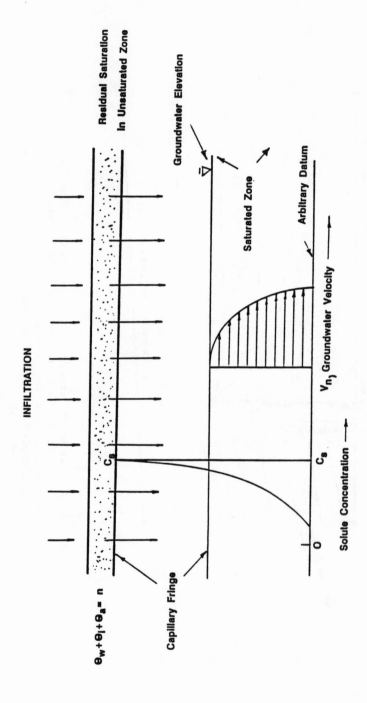

Figure J-5.   Immiscible Contaminant on Capillary Zone,
Density Less than Pure Water.

Infiltrating water, under draining conditions will reach an equilibrium with the immiscible liquid resulting in a saturated solute condition. For hypothetical purposes only, if it is assumed that only vertical groundwater flow exists in the capillary zone, the rate of solute input to the saturated zone will be limited by the rate of infiltration and C_s. If it assumed, again for illustrative purposes, that a horizontal flow boundary exists at the groundwater table, then mass transfer of solute from the capillary zone into the saturated zone will have only limited effects on the rate of solute input into the saturated zone. When considering the quantities of water infiltrating through the capillary zone per year in comparison to saturated flow rates, the above assumptions may be valid. The result is relatively, inefficient transfer of solutes from the capillary fringe zone to the saturated zone. Thus, in this scenario, pump and treat systems may not be necessary to remove the immiscible contaminants and advective dispersive dilution may be adequate to protect groundwater resources. Without knowledge a priority of the immiscible liquid distribution and interaction with the capillary zone, and advective-dispersive transport characteristics at a site, this approach may be risky. An alternative, however, may be close monitoring of groundwater in the saturated zone near the spill area, as vapor extraction proceeds. If the scenario in Figure J-5 exists at a site then solute concentrations in groundwater will decrease with time and no pump, treat and skim system may be necessary, to achieve desired levels of remediating in soil and groundwater. If near field transport of solutes from the spill area increases steadily with time, then groundwater pumping may be necessary to employ at that time.

In the case of an immiscible contaminant with a density less than water with impingement on the saturated zone by penetration of the capillary zone, the potential for solute transfer from the unsaturated zone to the saturated zone is greatly increased. This scenario may result from the depression of the capillary zone in a spill event where considerable quantities of an immiscible contaminant are spilled, such as that shown in Figure J-3. Alternatively, fluctuating groundwater tables may result from a rise in the groundwater table through wetting (imbibition) of the capillary zone as described by characteristic curves for a given porous media and immiscible contaminant. Remembering that immiscible contaminants become immobilized once at residual saturation, the result of wetting the capillary zone may result in the condition shown in Figure J-6. The net result of this scenario is that saturated solute concentrations exists at the top of the horizontal flow zone of the saturated zone. This boundary condition enables substantially greater mass transfer of solute into the saturated zone, principally resulting from the upper flow boundary being the immiscible contaminant itself. In Figure J-5, the upper boundary was only solute at less than C_s, and solubilization was limited to that achieved through infiltration. Clearly, the difference in these two situations greatly affects the rate of solute input into the saturated zone and should affect remedial action responses. Unless the solute transport phenomena from an immiscible phase into the saturated zone is understood and physically defined at a site, then neither optimal remediation systems can be designed nor saturated zone solute transport models can be effectively utilized to predict the impact of immiscible liquid remediation on saturated zone solute transport.

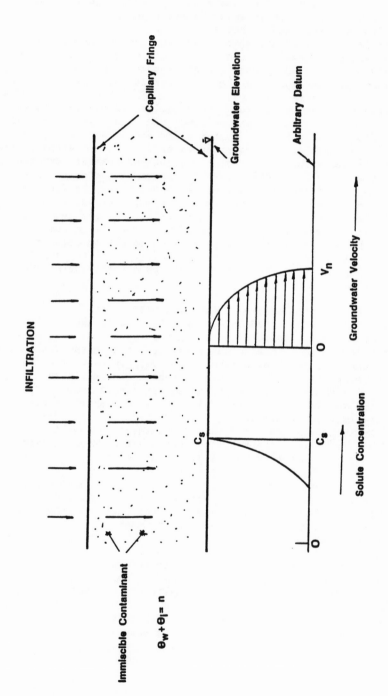

Figure J-6. Generalized Subsurface Condition of Immiscible Contaminant with Density Less Than Water Impinging on the Water Table.

Immiscible phase boundary conditions presented in Figures J-5 and J-6 also greatly effect the vapor diffusive flux rates from the capillary zone into the unsaturated zone. Bruell (1987) and Bruell and Hoag (1986) rigorously investigated the effect of immiscible liquid phase boundary conditions on subsequent hydrocarbon diffusive flux rates of benzene. For a given column geometry (diffusive path length of 47.6 cm), diffusive flux rates for benzene at 20 °C for an immiscible phase boundary condition similar to that shown in Figure J-5, resulted in benzene diffusive flux rates of 24.9 mg/cm^2-min and 6.1 mg/cm^2-min, for dry and wet (i.e. field capacity moisture content) concrete sand, respectively. Thus, moisture content played a significant role in reducing the effective porosity of the concrete sand. When residual saturation immiscible liquid phase boundary conditions were investigated the maximum benzene diffusive flux rate was 26.6 mg/cm^2-min, however the diffusive path length was only 22.4 cm. The moisture content in the residually saturated zone was 3.2 percent(v/v). When capillary zone immiscible liquid phase boundary conditions were investigated, the maximum benzene diffusive flux rate was reduced to 4.8 mg/cm^2-min with a diffusive path length of 22.4 cm. The moisture content in the capillary zone reflected saturated conditions (i.e., $\theta_v = n$). This research demonstrated that the immiscible liquid phase boundary condition greatly affects the diffusive flux rates of hydrocarbons that occur in the unsaturated zone. As the moisture content increases, then the diffusive flux rates of contaminants will decrease. The net result of these boundary conditions affects the concentrations of hydrocarbon vapors detected using soil gas assessment techniques and the rates of hydrocarbon recovery utilizing in situ vapor extraction.

With reference to Figure J-6 and knowing that advective air flowrates also decrease with increasing moisture content, creates a circumstance in the area of the capillary zone where advective air flow may not be in direct contact with the immiscible phase. Thus, diffusion in this case, will be the controlling mechanism of contaminant removal during vapor extraction. In the case depicted in Figure J-5, it is likely that some advective air flow will contact the immiscible phase, greatly increasing vapor extraction efficiency.

In the case of an immiscible liquid with a density greater than that of water, contaminant distribution is significantly different given a hypothetical spill to the subsurface. Penetration of the capillary and saturated zones by the immiscible liquid is likely, given sufficient spill volumes as shown is Figure J-7. Of great importance is the occurrence of groundwater flow through the immiscible liquid phase in the saturated zone, resulting in substantially greater solubilization rates of the immiscible phase and greater groundwater contamination potential than in the cases presented in Figures J-5 and J-6.

A typical in situ remedial action response to the dense immiscible liquid phase contamination is given in Figure J-8. Simultaneous vapor extraction and groundwater pumping are necessary to expose immiscible phase contaminants to advective air flow and to increase diffusive flux rates of contaminants in the vicinity of the groundwater table at time = T_2. In this case, dewatering of the saturated zone in the area of immiscible phase contaminants is desirable. Long-term plume management interceptor pumping

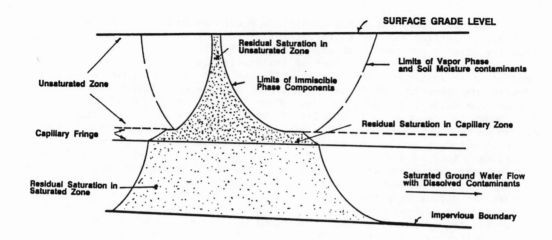

Figure J-7.   Generalized Subsurface Phase Distribution for Immiscible
Contaminants with Densities Greater than Pure Water.

strategies, such as those developed by Ahlfeld, Mulvey, Pinder and Wood (1988)
and Ahlfeld, Mulvey and Pinder (1988) should be implemented to optimally
circumvent uncontrolled groundwater contamination and to maximize groundwater
contaminant recovery rates. Strategies to maximize saturated zone dewatering
in the vicinity of the immiscible phase liquids must be developed to properly
implement this approach. Additionally, in situ bioremediation may be
considered as an additional technology to further degrade the immiscible
liquid, if complete subsurface dewatering is not possible.

SUMMARY

    Significant advances have been made in the past five years in the
understanding of volatile and semi-volatile contaminant behavior as related to
vapor extraction technologies. Coupled modeling of both contaminant behavior
and advective air flow, however remains limited to one dimensional systems.
Given the significant hydrogeological complexity of porous media and
subsequent heterogeneous distributions of immiscible phase contaminants, the
design utility of higher dimensional coupled models is questionable. Higher
dimensional advective air flow models are being used to design vapor
extraction systems. These models are generally dependent on site specific
parameters best determined in field air pumping tests, unless uniform
hydrogeologic conditions prevail with quantifiable boundary conditions
necessary for model design predictions. Three dimensional models are being

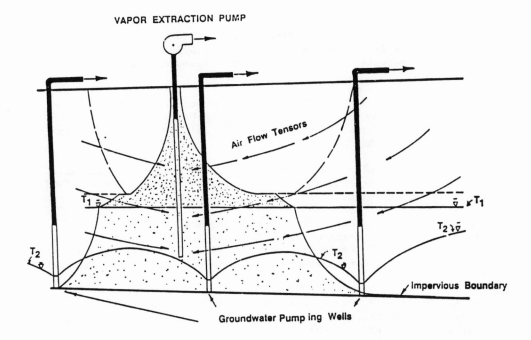

Figure J-8.   Typical In-Situ Remedial Actions to Immiscible Liquid Phase
Contaminants with Densities Greater than Water.

adapted to deal with non-radial symmetry and will be necessary to rigorously
model multiple extraction well and extraction well/injection well
applications.

Significant modeling and experimental research is needed to further
understand immiscible contaminant behavior in the capillary zone and adjacent
boundary conditions. The interaction of immiscible phase liquids in the
capillary zone with unsaturated zone infiltration and saturated zone transport
must be the focus of this research. The approach should include both
hydrological characteristics and testing procedures necessary to determine the
influencing factors. Chemical fate and transport in the unsaturated zone
under natural and advective air flow conditions must also be better understood
to wore effectively apply optimal in situ remediation processes.

Emphasis should be placed on basic research in the above areas, to be
followed at the appropriate time by demonstration level projects. When
demonstration level projects precede basic research needs, as has frequently
been the case in the past five years, the result generally do not properly
reflect necessary parameter control or monitoring and either inconclusive or
misleading results may be generated.

REFERENCES

Ahlfeld, D. P., Mulvey, J. M., and Pinder, G.F. 1988.Contaminated Groundwater Remediation Design Using Simulation, Optimization and Sensitivity Theory: 2, Analysis of Field Site. Water Resources Research. 24(3):443-452.

Ahlfeld, D. P., Mulvey, J. M., Pinder, G.F, and Wood, E. F.1988. Contaminated Groundwater Remediation Design Using Simulation, Optimization and Sensitivity Theory: 1, Model Development. Water Resources Research. 24(3):431-442.

Baehr, A. L. 1984. Immiscible Contaminant Transport in Soils with an Emphasis on Gasoline Hydrocarbons. Ph.D. Dissertation, Dept. of Civil Eng., University of Delaware.

Baehr, A. L., and Hoag, G. E. 1985. A Modeling and Experimental Investigation of Venting Gasoline Contaminated Soils. In: E. J. Calabrese and P. T. Kostecki, (editors), Soils Contaminated by Petroleum: Environmental and Public Health Effects, Wiley, New York, 458 pp.

Baehr, A. L., Hoag, G. E., and Marley, M. C. 1988. Removing Volatile Contaminants from the Unsaturated Zone by Inducing Advective Air-Phase Transport. Journal of Contaminant Hydrology, 4:1-26.

Bruell, C. J. 1987. The Diffusion of Gasoline-Range Hydrocarbons in Porous Media. Ph. D. Dissertation, Environmental Engineering, University of Connecticut, Storrs.

Bruell, C. J., and Hoag, G. E. 1987. The Diffusion of Gasoline-Range Hydrocarbon Vapors in Porous Media, Experimental Methodologies. Proceedings of Petroleum Hydrocarbons and Organic Chemicals in Ground Water: Prevention, Detection and Restoration. National Water Well Association and the American Petroleum Institute, Houston. 420-443.

Hoag, G. E., and Cliff, B. 1985. The Use of the Soil Venting Technique for the Remediation of Petroleum Contaminated Soils. In: E. J. Calabrese and P. T. Kostecki, (editors), Soils Contaminated by Petroleum: Environmental and Public Health Effects, Wiley, New York, 458 pp.

Lenhard, R. J., and Parker, J. C. 1986. Measurement and Prediction of Saturation-Pressure Relationships in Air-Organic Liquid-Water-Porous Media Systems. Virginia Poly. Tech. and State Univ.

Marley, M. C. 1985. Quantitative and Qualitative Analysis of Gasoline Fractions Stripped by Air from the Unsaturated Zone. M.S. Thesis, Department of Civil Engineering,the University of Connecticut, 87 pp.

Marley, M. C., and Hoag, G. E.   1984.   Induced Soil Venting for Recovery and
     Restoration of Gasoline Hydrocarbons in the Vadose Zone.   Proceedings of
     Petroleum Hydrocarbons and Organic Chemicals in Ground Water:
     Prevention, Detection and Restoration.   National Water Well Association
     and the American Petroleum Institute, Houston.   473-503.

Muskat, M., and Botset, H. G.   1931.   Flow of Gas Through Porous Materials.
     Physics.   1:27-47.

Parker, J. C., Lenhard, R. J., and Kuppusamy.   1986.   A Parametric Model for
     Constitutive Properties Governing Multiphase Fluid Conduction in Porous
     Media.   Virginia Poly. Tech. and State Univ.

Thorton, J. S., and Wootan, W. L.,   1982.   Venting for the Removal of
     Hydrocarbon Vapors from Gasoline Contaminated Soil.   J. Environ. Sci.
     Health, A17(1);31-44.

Wootan, W. L., and Voynick, T.   1984.   Forced Venting to Remove Gasoline Vapor
     From A Large-Scale Model Aquifer.   American Petroleum Institute.   82101-
     F:TAV.

Appendix K—State Regulatory Agency Contacts and Soil Cleanup Criteria

STATE	OFFICE	CONTACT	TELEPHONE	ALLOWED RESIDUAL
ALABAMA	DEPARTMENT OF ENVIRONMENTAL MANAGEMENT GROUNDWATER SECTION	SUSAN CHAMLES	(205)271-7832	SITE SPECIFIC--<1 ppm TPH
ALASKA	AIR & SOLID WASTE MANAGEMENT	JEFF MACH	(907)465-2653	NO STATE REGULATIONS
ARIZONA	UNDERGROUND STORAGE TANK COMPLIANCE	KIM MACAFEE	(602)257-2300	TPH - 100 ppm BENZENE - .130 ppm TOLUENE - 200 ppm ETHYL BENZENE - 68 ppm XYLENE -44 ppm
ARKANSAS	UNDERGROUND STORAGE TANKS	LINDA GRESCHEN	(501)562-7444	NO STATE REGULATIONS
CALIFORNIA	STATE WATER RESOURCES-- UNDERGROUND STORAGE TANK DIVISION	BETTY MORANO	(916)739-4436	BASED ON MODELS-- HIGHEST-- BENZENE--1 ppm EACH TEX--50 ppm TPH--1000 ppm LOWEST--TPH--10 ppm
COLORADO	DEPARTMENT OF ENVIRONMENTAL HEALTH	SCOTT WINTERS	(303)320-8333	NO STATE REGULATIONS
CONNECTICUT	OIL AND CHEMICAL SPILLS	JAMES SANPACROCE	(203)566-4633	NO MINIMUMS; CLEANUP TO LIMITS OF PROPERTY LINE; GROUNDWATER LIMITS
DELAWARE	UNDERGROUND STORAGE TANKS	EMIL ONUSCHAK	(302)323-4588	SITE BY SITE BTEX --1 ppm
DISTRICT OF COLUMBIA	EPA GUIDANCE	ANGELO TOMPROSE	(202)783-3194	100 ppm TPH
FLORIDA	BUREAU OF WASTE CLEANUP	TODD ALLEN	(904)487-3299	NO FORMAL REGULATIONS-- >500 ppm TPH-CHECK GROUND-WATER IMPACT <10 ppm NO CLEANUP REQUIRED <500 ppm THC <100 ppb TOTAL AROMATICS

APPENDIX K
STATE REGULATORY AGENCY CONTACTS AND SOIL CLEANUP CRITERIA
(continued)

STATE	OFFICE	CONTACT	TELEPHONE	ALLOWED RESIDUAL
GEORGIA	ENVIRONMENTAL PROTECTION DEPARTMENT-- INDUSTRIAL WASTE MANAGEMENT PROGRAM	CLIFF TRUSSELL	(404)669-3297	GASOLINE--SHOULD CLEAN TO 10 ppm--BTEX, IF BTEX >10 ppm, CLEAN TO BACKGROUND GENERALLY (PRACTICE NOT LAW)
HAWAII	UNDERGROUND STORAGE TANKS	LIZ ALBAZ	(808)548-8837	VISIBILITY, ODOR, SOME SOIL SAMPLING
IDAHO	DEPARTMENT OF ENVIRONMENT-WATER QUALITY BUREAU		(208)334-5839	RCRA RULES (HAZARDOUS) TO BACKGROUND IF FEASIBLE (NON-HAZARDOUS)
ILLINOIS	EPA-DIVISION OF LAND POLLUTION CONTROL	KELLY DUNBAR	(217)782-6761	BENZENE- 5 ug/kg TOLUENE- 2mg/kg ETHYL BENZENE-680ug/kg XYLENES- 1400 ug/kg
INDIANA	DEPARTMENT OF ENVIRONMENTAL MANAGEMENT	MANUELLA JOHNSON	(317)243-5060	SITE BY SITE TPH TEST FOR DETERMINATION >100 ppm CLEANUP IS REQUIRED BACKGROUND IF POSSIBLE
IOWA	DEPARTMENT OF NATURAL RESOURCES	JIM HORN	(515)281-8964	NO STATE REGULATIONS
KANSAS	DEPARTMENT OF HEALTH AND ENVIRONMENT	MARY JANE STELL	(913)296-1500	TPH--100 mg/kg BENZENE--1.4 mg/kg 1,2 DCE--8 mg/kg
KENTUCKY	DIVISION OF WASTE MANAGEMENT	DOUG BONK	(502)564-3382	NO STATE REGULATIONS
LOUISIANA	OFFICE OF SOLID AND HAZARDOUS WASTE MANAGEMENT	GEORGE GULLET	(504)342-7808	SITE BY SITE BTEX 50-100 ppm
MAINE	DEPT. OF ENVIRONMENTAL MANAGEMENT WASTE MANAGEMENT	RON SEVERENCE	(207)289-2651	NO STATE REGULATIONS -- SITE SPECIFIC
MARYLAND	DEPARTMENT OF ENVIRONMENT	HERB NEAD	(301)631-3442	TPH <100 ppm, IF LOW FLASHPOINT, MIX WITH CLEAN SOIL

APPENDIX K
STATE REGULATORY AGENCY CONTACTS AND SOIL CLEANUP CRITERIA
(continued)

STATE	OFFICE	CONTACT	TELEPHONE	ALLOWED RESIDUAL
MASSACHUSETTS	DEPT. OF ENVIRONMENTAL PROTECTION	CHARLIE TUTTLE	(617)292-5903	SITE SPECIFIC DEVELOPING NEW POLICY, AS OF (10/4/89)
MICHIGAN	LAND AND MANAGEMENT DEVELOPMENT	DON PARSONS	(517)373-1170	10 mg/kg TOTAL VOC WITH <15 % ABOVE 1 mg/kg
MINNESOTA	POLLUTION CONTROL AGENCY	SHIELA GROW	(612)297-2316	GUIDELINES--<10 ppm TPH, with ORGANIC VAPOR ANALYZER, OK;10-100ppm CLEAN UP NECESSARY,FUEL OIL CLEAN TO 1ppm OVER BACKGROUND
MISSISSIPPI	BUREAU OF POLLUTION CONTROL	JACK McCORD	(601)961-5062	1 COMPOSITE SAMPLE/TANK ACTION--100 ppm(mg/kg) <10 % LEL
MISSOURI	DEPARTMENT OF NATURAL RESOURCES	FRED HUTSON	(314)751-7326	SITE SPEC--GENERALLY <10 ppm TPH <1 ppm BTEX
MONTANA	SOLID AND HAZARDOUS WASTE BUREAU	DOUG ROGNESS	(406)444-3454	SITE SPECIFIC
NEBRASKA	DEPARTMENT OF ENVIRONMENTAL CONTROL	JIM BOROVICH	(402)471-4230	TPH 2 ppm BENZENE .005 ppm TOLUENE 2.42 ppm ETHYLBENZENE 1.4 ppm XYLENE 0.4 ppm
NEVADA	DIVISION OF UNDERGROUND STORAGE TANKS	ALAN BIAGGI	(702)885-5872	100 ppm TPH--*RCRA-IGNITIBILITY, LEAD CONTENT
NEW HAMPSHIRE	DEPARTMENT OF ENVIRONMENTAL SERVICES	TIM DREW	(603)271-3306	*TOTAL VOLATILES; LEAD
NEW JERSEY	DIVISION OF WATER RESOURCES	ANALAB, INC	(609)984-3156	BENZENE - 0.07 mg/l TOLUENE - 14.4 mg/l DCE - 0.1 mg/l

APPENDIX K
STATE REGULATORY AGENCY CONTACTS AND SOIL CLEANUP CRITERIA
(continued)

STATE	OFFICE	CONTACT	TELEPHONE	ALLOWED RESIDUAL
NEW MEXICO	DEPARTMENT OF HEALTH AND ENVIRONMENT ENVIRONMENTAL IMPACT DIVISION		(505)827-2894	VISUAL & OLFACTORY-- OBSERVE AFTER AERATION
NEW YORK	DEPARTMENT OF UNDERGROUND STORAGE TANKS	FRANK PEDUDO	(518)457-2462	SITE SPECIFIC - GENERALLY DRINKING WATER STANDARDS, WITH TOXICITY CHARACTERISTICS OF LEACHATE POTENTIAL TEST, IF <DW STDS, OK
NORTH CAROLINA	DEPARTMENT OF HEALTH AND HUMAN SERVICES	BILL JEETER	(919)733-5083	>100 ppm TPH REMEDIATE 10-100 ppm MONITOR <10 ppm CONSIDERED CLEAN THC < 100ppm
NORTH DAKOTA	HEALTH DEPARTMENT	GARY BERRETH	(701)224-2366	SITE SPECIFIC
OHIO	EPA OFFICE OF SOLID & HAZARDOUS WASTE MANAGEMENT	TOM FORBES	(614)752-7938	BTEX 2 ppb; TPH 1 ppm
OKLAHOMA	CORPORATION COMMISSION	TANA WALKER	(405)521-3107	ACTION LEVELS TPH-500 ppm BTEX-10 ppm
OREGON	DEPARTMENT OF ENVIRONMENTAL QUALITY	MIKE ANDERSON	(503)229-5731	SITE SPECIFIC MODELS LEVEL 1-TPH 40 ppm LEVEL 2-TPH 80 ppm LEVEL 3-TPH 130 ppm
PENNSYLVANIA	NON-POINT SOURCES AND STORAGE TANKS	JOHN BORLAND	(717)787-2666	NO STATE REGULATIONS

APPENDIX K
STATE REGULATORY AGENCY CONTACTS AND SOIL CLEANUP CRITERIA
(continued)

STATE	OFFICE	CONTACT	TELEPHONE	ALLOWED RESIDUAL
RHODE ISLAND		DAVID SHELDON	(401)277-2808	VISUAL OBSERVANCE-- SHIPPED TO STATE APPROVED FACILITY
SOUTH CAROLINA	DEPT OF HEALTH & ENVIRONMENTAL CONTROL	PRESTON CAMPBELL	(803)734-5331	THC-100 ppm
SOUTH DAKOTA	DEPT OF WATER AND NATURAL RESOURCES	DICK PIEFER	(605)773-3351	TPH-10 ppm
TENNESSEE	DEPT OF GROUNDWATER PROTECTION	DON GILMORE	(615)741-4094	10 ppm BTEX (GASOLINE) 100 ppm TPH
TEXAS	WATER COMMISSION AND DEPT OF HEALTH CORRECTIVE ACTION	DAN AIREY	(512)463-7972	<500 ppm BTX ; IGNITIBILITY
UTAH	BUREAU OF SOLID AND HAZARDOUS WASTE	BOB FORD	(801)538-6121	SITE SPECIFIC 100 mg/l THC IN SOIL
VERMONT	AGENCY OF NATURAL RESOURCES	PAUL VAN HOLIBEKE	(802)244-5674	SITE SPECIFIC-SOIL MAY BE PLACED ON SITE IF <20 ppm AS BENZENE
VIRGINIA	STATE WATER CONTROL BOARD	STEVE WILLIAMS	(804)367-0970	SITE SPECIFIC THROUGH RISK ASSESSMENT 100 ppm (GENERALLY)
WASHINGTON	ECOLOGY DEPT	JOE HICKEY	(206)867-7000	TPH 200 ppm BENZENE 660 ppb TOLUENE 143 ppm ETHYLBENZENE 14 ppm XYLENE 900 ppm
WEST VIRGINIA	UNDERGROUND STORAGE TANKS	PAT BOYD	(304)348-5935	BACKGROUND
WISCONSIN	BUREAU OF SOLID & HAZARDOUS WASTE MGMT		(608)266-1327	SITE SPECIFIC--10-50 ppm THC OR LOWER DEPENDING ON GROUNDWATER LEVELS
WYOMING	DEPARTMENT OF ENVIRONMENTAL QUALITY	DAVE MONTAGUE	(307)777-7781	NO STATE REGULATIONS OLFACTORY LEVELS

APPENDIX K

STATE REGULATORY AGENCY CONTACTS AND SOIL CLEANUP CRITERIA

(continued)

LEGEND:

REFERENCE: EPA SURVEY OF STATE PROGRAMS PERTAINING TO CONTAMINATED SOILS (1988)

TPH - TOTAL PETROLEUM HYDROCARBONS

BTEX - BENZENE, TOLUENE, ETHYLBENZENE, XYLENE

LEL - LOWER EXPLOSIVE LIMIT

TEX - TOLUENE, ETHYLBENZENE, XYLENE

VOC - VOLATILE ORGANIC COMPOUND

GW - GROUNDWATER

THC - TOTAL HYDROCARBON

DCE - DICHLOROETHYLENE

RCRA - RESOURCE CONSERVATION RECOVERY ACT

Appendix L—State Regulatory Agency Contacts and Air Discharge Criteria

STATE	OFFICE	CONTACT	TELEPHONE	AIR QUALITY EMISSIONS
ALABAMA	AIR QUALITY	TIM OWENS	(205)271-7861	EMISSION OF AIR TOXICS >0.1 lbs/hr STATE REQUIRES MORE INFORMATION ON SOURCE
ALASKA	AIR & SOLID WASTE MANAGEMENT	JOHN SANSTEND	(907)465-2666	40 CFR 60
ARIZONA	AIR QUALITY	CARROLL DEKLE	(602)257-2300	40 CFR 60
ARKANSAS	AIR QUALITY	J.B. JONES	(501)562-7444	NO STANDARDS
CALIFORNIA	AIR RESOURCES BOARD	BOB FLETCHER	(916)739-8267	EACH COUNTY HAS OWN
COLORADO	AIR POLLUTION	JOHN PLOG	(303)331-8500	REASONABLE CONTROL TECHNOLOGY
CONNECTICUT	AIR COMPLIANCE	JOHN GOVE	(203)566-2690	COMPOUND SPECIFIC per 8 hr average
DELAWARE	UNDERGROUND STORAGE TANKS	EMIL ONUSCHAK	(302)323-4588	2.4 lbs/day of VOCs
DISTRICT OF COLUMBIA	AIR MONITORING	DON WAMBSGANS	(202)767-7370	1 lb/day of VOCs
FLORIDA	AIR QUALITY	BARRY ANDREWS	(904)488-1344	GUIDANCE POLICY-MAXIMUM IMPACT vs. ACCEPTED FRACTION OF TLV
GEORGIA	AIR PROTECTION	LYNN RHODES	(404)656-6900	PERMIT REQUIRED FOR ALL EMISSIONS; IF PROCESS IS IN-CINERATION, 90% OXIDATION REQUIRED
HAWAII	DEPARTMENT OF HEALTH POLLUTION CONTROL BRANCH	TYLER SUGIHARA	(808)543-8205	NSPS - 40 CFR 60
IDAHO	ENVIRONMENTAL QUALITY	MARTIN BAUER	(208)334-5834	NO LIMITS - SITE SPECIFIC
ILLINOIS	AIR QUALITY PLANNING	JOHN REED	(217)782-7326	VOCs CONSIDERED PHOTOCHEMICAL-LY REACTIVE (RULE 66) - 8 lb/hr or 85% CONTROL

APPENDIX L
STATE REGULATORY AGENCY CONTACTS AND AIR DISCHARGE CRITERIA
(Continued)

STATE	OFFICE	CONTACT	TELEPHONE	AIR QUALITY EMISSIONS
INDIANA	AIR QUALITY STANDARDS AND PLANNING	ANN BLACK	(317)247-5110	SITE BY SITE - MUST REGISTER
IOWA	DEPARTMENT OF NATURAL RESOURCES	REX WALKER	(515)281-5145	IF CARCINOGEN INVOLVED AND THE MAXIMUM CONCENTRATION AT GROUND LEVEL >10(-6)
KANSAS	AIR EMISSIONS	GENE SALLEE	(913)296-1500	MODELED FOR BENZENE REPORTING REQUIREMENTS IF >10 TONS EMISSION/YEAR PERMITS
KENTUCKY	AIR QUALITY	MARJORIE MULLIN	(502)564-3382	>85% REMOVAL OF INPUT BY WEIGHT
LOUISIANA	AIR PERMITS	KARL OSTERTAHLER	(504)342-9047	NO STANDARDS - MAKE A MODEL & COMPARE TO 1/42 OF TLV
MAINE	AIR QUALITY	FRED LAVALEE	(207)289-2437	NO REQUIREMENTS
MARYLAND	AIR MANAGEMENT	BONNIE WATTS	(301)631-3285	PERMITS REQUIRED
MASSACHUSETTS	AIR QUALITY	RICH DRISCOLL	(617)292-6630	MODELED; PERMITS REQUIRED
MICHIGAN	AIR QUALITY	STEVE KISH	(517)373-7023	PCE - 0.0024 PPM TCE - 0.0073 PPM
MINNESOTA	AIR STANDARDS	AHTO NIEAMIEJA	(612)296-7802	NO EMISSIONS STANDARDS
MISSISSIPPI	AIR QUALITY	JACK McCORD	(601)961-5171	COMPLY WITH FEDERAL STANDARDS
MISSOURI	PLANNING	TODD CRAWFORD	(314)751-7929	NATIONAL EMISSIONS STANDARD HAZARDOUS AIR POLLUTANTS (NESHAPS)
MONTANA	AIR QUALITY	BOB RAUSCH	(406)444-2821	NO STANDARDS
NEBRASKA	AIR QUALITY	JOE FRANCIS	(402)471-2189	REPORTING REQUIRED IF >15 lbs/hr OR 100 lbs/day VOCs

APPENDIX L

STATE REGULATORY AGENCY CONTACTS AND AIR DISCHARGE CRITERIA

(Continued)

STATE	OFFICE	CONTACT	TELEPHONE	AIR QUALITY EMISSIONS
NEVADA	AIR QUALITY	BAY McCLEARY	(702)885-5065	NO STATEWIDE REGULATIONS
NEW HAMPSHIRE	AIR QUALITY	ANDY BARDENARIK	(603)271-1370	ACTION LEVELS TOLUENE - 0.8 ppm ETHYLBENZENE - 0.1 ppm XYLENE - 0.55 ppm TPH as BENZENE 1 ppm/24 hr average At Nearest Public Receptor
NEW JERSEY	UNDERGROUND STORAGE TANKS	DIANE PUPA	(609)292-6383	MAXIMUM EMISSION RATE IS 15% OF INPUT BY WEIGHT @ 7 lbs/hr SOME GREATER RESTRICTIONS OCCUR FOR HIGHER CONCENTRATIONS SEE NJ TITLE 7:CH 27: SUB 16
NEW MEXICO	ENVIRONMENTAL IMPROVEMENT BUREAU	BOB KIRKPATRICK	(505)827-0070	0.19 ppm TPH/3hr Average FOR SCREENING SITES ONLY
NEW YORK	UNDERGROUND STORAGE TANKS	ANTHONY KARWIL	(518)457-2462	NO STATE REGULATIONS
NORTH CAROLINA	AIR QUALITY	BILL JEETER	(919)733-5083	40 lbs/day VOCs THEN IT DEPENDS ON REGIONAL OFFICES
NORTH DAKOTA	AIR QUALITY	TOM BACHMAN	(701)224-2348	NO SPECIFIC REGULATIONS
OHIO	AIR POLLUTION CONTROL MANAGEMENT	BILL JURIS	(614)644-2270	8 lbs/hr, 40 lbs/day VOC PERMITS NECESSARY
OKLAHOMA	AIR QUALITY	LARRY BYRUM	(405)271-5220	REGULATIONS FOR VOCs ARE BEING CHANGED AS OF 10/4/89
OREGON	AIR QUALITY	RAY POTTS	(503)229-6411	NO STANDARDS-PERMITS REQUIRED IF >70 TONS YEAR EMITTED
PENNSYLVANIA	AIR QUALITY	JOHN CLARKE	(717)783-9248	ODOR STANDARDS @ PROPERTY LINES

APPENDIX L
STATE REGULATORY AGENCY CONTACTS AND AIR DISCHARGE CRITERIA
(Continued)

STATE	OFFICE	CONTACT	TELEPHONE	AIR QUALITY EMISSIONS
RHODE ISLAND	AIR & HAZARDOUS MATERIALS	CHRIS JOHN	(401)277-2808	EPA TOXICITY LEVELS
SOUTH CAROLINA	AIR QUALITY PERMITS	PRESTON CAMPBELL	(803)734-4541	NO EMISSIONS STANDARDS
SOUTH DAKOTA	AIR POLLUTION CONTROL	TIM ROGERS	(605)773-3151	NO EMISSIONS STANDARDS
TENNESSEE	AIR POLLUTION CONTROL	BILL CLELAND	(615)741-3651	SITE-SPECIFIC - NEED PERMITS
TEXAS	AIR QUALITY CORRECTIVE ACTION	KEITH COPELAND	(512)463-7786	TPH 100 ppm TOTAL BTEX 30 ppm
UTAH	AIR STANDARDS	JEFF MINUS	(801)538-6108	<1 TON HYDROCARBONS/SITE
VERMONT	HAZARDOUS MATERIALS	CHUCK SCHMER	(802)244-5674	90% REMOVAL PERFORMANCE STANDARD
VIRGINIA	AIR QUALITY	BILL SYDNOR	(804)367-0970	NO STATEWIDE REGULATIONS
WASHINGTON	AIR QUALITY	DAN HOVAC	(206)867-7100	NO STATEWIDE REGULATIONS CONTACT LOCALITIES
WEST VIRGINIA	AIR POLLUTION CONTROL	BOB WEISER	(304)348-4022	40 CFR 60 - NSPS
WISCONSIN	AIR AND WASTE MANAGEMENT	LARRY BRUSS	(608)266-7718	3 lbs/hr; 15 lbs/day VOCs
WYOMING	AIR QUALITY	BERNIE DAILEY	(307)777-7391	40 CFR 60 -NSPS

LEGEND:
VOC - VOLATILE ORGANIC COMPOUNDS
40 CFR 60 - CODE OF FEDERAL REGULATIONS, Vol. 40, Part 60
TLV- THRESHOLD LIMIT VALUE
NSPS - NEW SOURCE PERFORMANCE STANDARDS
PCE - PERCHLOROETHYLENE
TCE - TRICHLOROETHYLENE
UST - UNDERGROUND STORAGE TANKS

Glossary

Adsorption. The attraction of ions or compounds to the surface of a solid.

Advection. The process of transfer of fluids (vapors or liquids) through a geologic formation in response to a pressure gradient that may be caused by changes in barometric pressure, water table levels, wind fluctuations or rainfall percolation.

Aerobic. In the presence of oxygen.

Air Permeability. A measure of the ability of a soil to transmit gases. It relates the pressure gradient to the flow.

Air/Water Separator. A device to separate, through additional retention time, physical means, or cooling, entrained liquids from a vapor stream.

Aliphatic. Of or pertaining to a broad category of carbon compounds distinguished by a straight, or branched, open chain arrangement of the constituent carbon atoms. The carbon-carbon bonds may be saturated or unsaturated.

Anaerobic. In the absence of oxygen.

Anisotropy. The dependence of property upon direction of measurement (e.g., hydraulic conductivity, porosity, compressibility, dispersion, etc.).

Aromatic. Of or pertaining to organic compounds that resemble benzene in chemical behavior.

Bentonite. A colloidal clay, largely made up of the mineral sodium montmorillonite, a hydrated aluminum silicate.

Biodegradation. A process by which microbial organisms transform or alter through enzymatic action the structure of chemicals introduced into the environment.

Bulk Density. The oven-dried mass per unit volume (including pore space) of soil.

Capillary Fringe. The zone of a soil above the water table within which most of the soil is saturated, but is at less than atmospheric pressure. The capillary fringe is considered to be part of the vadose zone but not of the unsaturated zone.

Catalyst. A substance that increases the rate of a chemical reaction and may be recovered essentially unaltered in form and amount at the end of the reaction.

Coarse-Textured Soils. Soils comprised primarily of particles with relatively large diameters (e.g., sand, loam).

Darcy's Law. An empirical relationship between hydraulic gradient and the viscous flow of water in the saturated zone of a porous medium under conditions of laminar flow. The flux of vapors through the voids of the vadose zone can be related to pressure gradient through the air permeability by Darcy's Law. See hydraulic conductivity, air permeability, hydraulic gradient, pressure gradient, laminar flow, vadose zone, saturated zone.

Density. The amount of mass per unit volume.

Diffusion. The process by which molecules in a single phase equilibrate to a zero concentration gradient by random molecular motion. The flux of molecules is from regions of high concentration to low concentration and is governed by Fick's Second Law.

Dispersion. The process by which a substance or chemical spreads and dilutes in flowing groundwater or soil gas.

Emissions Control Device. The equipment used to remove pollutants from the exhaust stream of a soil vapor extraction system.

Enhanced Biotreatment. The phenomenon sometimes noticed after SVE whereby naturally-occurring soil microbes degrade the soil contaminant at an increased rate, perhaps due to higher available soil oxygen.

Entrainment. A process in which suspended droplets of liquid are carried in the vapor stream.

Fick's Second Law. An equation relating the change of concentration with time due to diffusion to the change in concentration gradient with distance from the source of concentration. See diffusion, effective diffusion coefficient.

Field Capacity. The percentage of water remaining in the soil 2 or 3 days after gravity drainage has ceased from saturated conditions.

Fine Textured Soils. Soils comprised primarily of particles with small diameters (e.g., silt, clay).

Flow Lines. Lines on a cross-sectional diagram that show the direction of flow of air through the soil. Flow lines are used in the design of vapor extraction systems.

Flux. The rate of movement of mass through a unit cross-sectional area per unit time in response to a concentration gradient or some advective force, having units of mass per area per time (g/cm^2-sec).

Free Product. A contaminant in the unweathered phase, where no dissolution or
    biodegradation has occurred. See non-aqueous phase liquid.

Fugacity. Escaping tendency of a chemical substance from a particular phase.
    Fugacity is measured in units of pressure, and the higher the fugacity
    the greater tendency of chemical to escape from a phase.

Gas Chromatography. The process in which the components of a mixture are
    separated from one another by volatilizing the sample into a carrier gas
    stream. Different components move through a bed of packing or a coated
    capillary tube at different rates, and so appear one after another at
    the effluent end, where they are detected and measured by thermal
    conductivity changes, density differences, or ionization detectors.

Gasoline. A mixture of volatile hydrocarbons suitable for use in a
    spark-ignited internal engine and having an octane number of at least
    60. The major components are branched-chained paraffins,
    cyclaparaffins, and aromatics.

Henry's Law. The relationship between the partial pressure of a compound and
    the equilibrium concentration in the liquid through a constant of
    proportionality known as Henry's Law Constant. See partial pressure.

Heterogeneity. The dependence of property upon location of measurement (e.g.,
    hydraulic conductivity, porosity, compressibility, dispersion, etc.).
    Heterogeneity may be due to grain size trends, stratigraphic contacts,
    faults, and vertical bedding.

Homogeneity. The independence of property with location of measurement.

Hydraulic Conductivity. The constant of proportionality in Darcy's Law
    relating the rate of flow of water through a cross-section of porous
    medium in response to a hydraulic gradient. Also known as the
    coefficient of permeability, hydraulic conductivity is a function of the
    intrinsic permeability of a porous medium and the kinematic viscosity of
    the water which flows through it. Hydraulic conductivity has units of
    length per time (cm/sec).

Hydrocarbon. Any of a large group of compounds composed only of carbon and
    hydrogen.

Hysteresis. The dependence of the state of a system on direction of the
    process leading to it; a non-unique response of a system to stress,
    responding differently when the stress is released. Compressibility,
    moisture content, soil adsorption and unsaturated hydraulic conductivity
    exhibit hysteretic behavior.

Immiscible. Incapable of being mixed, such as oil and water.

Infiltration. The downward movement of water through a soil from rainfall or
    the application of artificial recharge in response to gravity and
    capillarity.

Injection Well. A well used during soil vapor extraction into which air is forced under pressure.

Inlet Well. A well used during soil vapor extraction through which air is allowed to enter the soil passively.

Isopotential lines. Lines indicating areas of equal pressure. On a cross-sectional diagram in an isotropic medium, isopotential lines intersect flow lines at right angles.

Isotherm. A relationship between the amount of solute adsorbed (expressed as a mass percentage of adsorbent) and the concentration in the influent vapor stream, at a given temperature and pressure.

Isotropy. The independence of a property with direction of measurement (e.g., hydraulic conductivity, porosity, compressibility, dispersion, etc.).

Lower Explosive Limit. The concentration of a gas below which the level is insufficient to support an explosion.

Microorganisms. Microscopic organisms including bacteria, protozoans, yeast fungi, viruses and algae.

Macropore. A large pore in a porous medium which may be formed by physical phenomena or biological activity, and through which water, or other fluids, flows solely under the influence of gravity, unaffected by capillarity.

Moisture Content. The amount of water lost from the soil upon drying to a constant weight, expressed as the weight per unit weight of dry soil or as the volume of water per unit bulk volume of the soil. For a fully saturated medium, moisture content equals the porosity; in the vadose zone, moisture content ranges between zero and the porosity value for the medium. See porosity, vadose zone, saturated zone.

Mottling. The reticulate coloring pattern of soils which is indicative of alternating oxidizing/reducing conditions. Mottling implies seasonally wet soil conditions.

NAPL. Non-aqueous phase liquid.

Oxidation. A chemical reaction that increases the oxygen content of a compound, or raises the oxidation state of an element.

Oxidation Potential. The difference in potential between an atom or ion and the state in which an electron has been removed to an infinite distance from this atom or ion.

Partial Pressure. The portion of total vapor pressure due to one or more constituents in a vapor mixture.

Particle Density. The amount of mass of a substance per unit volume of the substance.

Percolation. The downward movement of water through soil. Especially, the downward flow of water in saturated or nearly saturated soil at hydraulic gradients of the order of 1.0 or less.

Permeability. A measure of a soil's resistance to fluid flow. Permeability, along with fluid viscosity and density, is used to determine fluid conductivity.

Piezometer. An instrument used to measure pressure. Often used in reference to tubes inserted into the soil for measuring soil pressure or water table depth.

Porosity. The volume fraction of a rock or unconsolidated sediment not occupied by solid material but usually occupied by water and/or air. Porosity is a dimensionless quantity that is expressed as a percent or a decimal.

Pressure Gradient. A pressure differential in a given medium, such as water or air, which tends to induce movement from areas of higher pressure to areas of lower pressure.

Pulsed Venting. A method of operation in which the vacuum pump or blower is operated intermittently. During periods when the vacuum is off, the contaminant vapors re-equilibrate. When the system is turned back on, extracted vapors have higher concentrations. Pulsed venting is cheaper than continuous venting due to lower power consumption.

Radius of Influence. The maximum distance away from a vacuum source that is still affected by the vacuum.

Residual Saturation. The amount of water or oil remaining in the voids of a porous medium and held in an immobile state by capillarity and dead-end pores.

Saturated Zone. The zone of the soil below the water table where all space between the soil particles is occupied by water.

Short Circuiting. As it applies to SVE, the entry of ambient air into the extraction well without passing through the contaminated zone by, for example, entering a utility trench.

Site Characterization. Documentation of all site characteristics that may impact the design of a subsurface venting system.

Soil Gas Survey. Investigation of the distribution of soil gas concentrations in three dimensions. The term may apply to the map or to data documenting the soil gas concentrations.

Soil Sorption Coefficient. A measure of the preference of an organic chemical to leave the dissolved aqueous phase in the soil and become attached or adsorbed to soil particles and organic carbon.

Soil Texture. Refers to the particle size distribution of the soil. Particle size classes of the USDA system depend on the relative proportions of sand (2-0.05 mm), silt (0.05-0.002 mm), and clay particles (<0.002 mm).

Soil Vapor Extraction. A soil remediation technique that involves removing contaminant-laden vapors from the soil under a vacuum. Also known as vacuum extraction, soil venting, soil stripping, and enhanced volatilization.

Solubility. The amount of mass of a compound that will dissolve into a unit volume of solution.

Sorption. A general term used to encompass the process of absorption, adsorption, ion exchange, and chemisorption.

TPH. See total petroleum hydrocarbons.

Total Petroleum Hydrocarbons. A measure of the mass or concentration of all the petroleum constituents present in a given amount of air, soil, or water.

Tortuosity. The ratio of path length through a porous medium to the straight-line flow path which describes the geometry of the porous medium. Tortuosity is a dimensionless parameter which ranges value from 1 to 2.

Unsaturated Zone. The portion of a porous medium, usually above the water table in an unconfined aquifer, within which the moisture content is less than saturation and the capillary pressure is less than atmospheric pressure. The unsaturated zone does not include the capillary fringe.

Upper Explosive Limit. The concentration of a gas above which the gas will not explode.

UST. See underground storage tank.

Underground Storage Tank. By statutory definition, any tank that is used to contain an accumulation of regulated product [basically, petroleum or hazardous substances] and the volume of which is 10 percent or more underground.

Vacuum. The existence of below-atmospheric pressure.

Vadose Zone. The portion of a porous medium above the water table within which the capillary pressure is less than atmospheric and the moisture content is less than saturation. The vadose zone includes the capillary fringe.

Vapor Density.   The amount of mass of a vapor per unit volume of the vapor.

Vapor Pressure.   The equilibrium pressure exerted on the atmosphere by a
    liquid or solid at a given temperature.   Also, a measure of a
    substance's propensity to evaporate or give off flammable vapors.   The
    higher the vapor pressure, the more volatile the substance.

Volatilization.   The process of transfer of a chemical from the water or
    liquid phase to the air phase.   Solubility, molecular weight, and vapor
    pressure of the liquid and the nature of the air-liquid/water interface
    affect the rate of volatilization.   See solubility, vapor pressure.

Water Content.   See moisture content.

Water Table.   The water surface in an unconfined aquifer at which the fluid
    pressure in the voids is at atmospheric pressure.

Weathering.   The process where a complex compound is reduced to its simpler
    component parts, transported through physical processes, or biodegraded
    over time.

Well Screen.   The segment of well casing which has slots to permit the flow of
    liquid or air but prevent the passage of soil or backfill particles.

DETECTION OF SUBSURFACE HAZARDOUS WASTE CONTAINERS BY NONDESTRUCTIVE TECHNIQUES

by

Arthur E. Lord, Jr. and Robert M. Koerner
Drexel University

Pollution Technology Review No. 172

This book describes the detection of subsurface hazardous waste containers by nondestructive testing (NDT) (remote-sensing) techniques.

There is a vast amount of hazardous waste buried below the surface of the soil. It is most important to clean up these wastes before they do (additional) damage to the environment. The first step in any cleanup procedure is to detect the waste and then determine its spatial extent. As in any subsurface exploration, many techniques can be brought to bear. Test borings and limited excavations are very valuable but are not without their problems. The information so obtained is not continuous and the destructive nature of the test makes it possible that waste could inadvertently be released during the probing phase. Therefore, there is an interest in probing from the surface with nonintrusive methods.

This study was undertaken with the goal of identifying and assessing the best possible NDT techniques for detecting and delineating hazardous waste. The work concentrated on the detection of steel and plastic containers buried beneath the surface of soil and water bodies. Seventeen techniques were considered and four were ultimately decided upon. They were: electromagnetic induction (EMI); metal detection (MD); magnetometer (MAG); and ground penetrating radar (GPR). The containers, both steel and plastic, varying in size from 5 gal to 55 gal, were buried in known distributions in a wide variety of soils; also, some were submerged in water. Five diverse field sites were used.

As a result of the work at the five field sites, a relatively complete picture has emerged concerning the strengths and weaknesses of the NDT container location techniques. Application of signal enhancement techniques (background suppression) can be expected to further improve NDT utility.

The listing below includes **chapter titles and selected subtitles.**

1. INTRODUCTION

2. CONCLUSIONS

3. RECOMMENDATIONS

4. POSSIBLE NDT METHODS WHICH CAN BE USED TO HELP SOLVE HAZARDOUS MATERIALS WASTE PROBLEMS
Microwaves (Pulsed)
Microwaves (Continuous)
Eddy Current
Magnetometer
Seismic Reflection
Seismic Refraction
Electrical Resistivity
Penetrating Radiation
Acoustic Emission
Liquid Penetrant
Infrared Radiometry
Pulse-Echo Ultrasonics
Sonar
Electromagnetic Induction
Induced Polarization
Self-Potential
Optical Techniques

5. DETAILS OF LITERATURE SEARCHES ON EACH NDT METHOD

6. EXPERIMENTAL EVALUATION OF THE DESIGNATED NDT TECHNIQUES AT VARIOUS FIELD SITES
Evaluation of the NDT Techniques at a Dry Sandy Soil Site with Little Interference ("Best Possible" Site)
Evaluation of the Four NDT Techniques at a Site with Saturated Silty Clay Soil
Evaluation of the Four NDT Techniques with Containers Located Beneath Water Surfaces
Evaluation of the Four NDT Techniques in a Saline Sandy Soil
Evaluation of the Four NDT Techniques at a Sandy Soil Site with Buried Plastic Containers

REFERENCES

APPENDICES

ISBN 0-8155-1224-4 (1990) $39   7" x 10"   83 pages